AF576035

Advances in Industrial and Environmental Microbiology

Advances in Industrial and Environmental Microbiology

Editors

Slawomir Ciesielski
Ivone Vaz-Moreira

MDPI • Basel • Beijing • Wuhan • Barcelona • Belgrade • Manchester • Tokyo • Cluj • Tianjin

Editors
Slawomir Ciesielski
University of Warmia and Mazury in Olsztyn
Poland

Ivone Vaz-Moreira
Universidade Catolica Portuguesa
Portugal

Editorial Office
MDPI
St. Alban-Anlage 66
4052 Basel, Switzerland

This is a reprint of articles from the Special Issue published online in the open access journal *Applied Sciences* (ISSN 2076-3417) (available at: https://www.mdpi.com/journal/applsci/special_issues/industrial_environmental_microbiology).

For citation purposes, cite each article independently as indicated on the article page online and as indicated below:

LastName, A.A.; LastName, B.B.; LastName, C.C. Article Title. *Journal Name* **Year**, *Volume Number*, Page Range.

ISBN 978-3-0365-2484-9 (Hbk)
ISBN 978-3-0365-2485-6 (PDF)

Contents

About the Editors

Slawomir Ciesielski Slawomir Ciesielski is a Professor of biology at the University of Warmia and Mazury in Olsztyn, Poland. He obtained habilitation degree in biotechnology from Wrocław University of Environmental and Life Sciences in 2011. He is a multidisciplinary scientist with a major interest in molecular microbiology. Currently his research is focusing on microbiological aspects of wastewater treatment, microbial production of polymers and gut microbiomes. Reviewer in national and international funding programs and relevant JCR-indexed journals. He has guided a number of students toward masters and doctoral degrees. He has participated as a researcher and principal investigator in several research projects and coordinator in national and international scientific activities.

Ivone Vaz-Moreira obtained a B.Sc. degree in Biochemistry from the University of Coimbra, Coimbra, Portugal in 2005 and a Ph.D. in Biotechnology with a specialization in Microbiology from the Portuguese Catholic University, Porto, Portugal in 2012. Currently she is a Researcher at the Portuguese Catholic University, Porto, Portugal working on bacterial ecology and antibiotic resistance in the environment. Her research is focused in Environmental Microbiology, associated with aquatic habitats and their intersections with soil/sediments, plants, animals, and, finally, humans. This includes the study of i) the bacterial diversity, in particular, on the urban water cycle and, ii) the potential of antibiotic resistance dissemination from aquatic environments. Collaborated in many taxonomy studies, having been considered one of the ten bacterial taxonomists in Portugal in 2014. Co-author of 50 Papers in Scientific Journals included in Science Citation Index (>2250 citations and h-index 24), and 15 book chapters.

Editorial

Advances in Industrial and Environmental Microbiology

Slawomir Ciesielski [1,*] and Ivone Vaz-Moreira [2,*]

1 Department of Environmental Biotechnology, University of Warmia and Mazury in Olsztyn, Sloneczna 45G, 10-709 Olsztyn, Poland
2 CBQF-Centro de Biotecnologia e Química Fina-Laboratório Associado, Escola Superior de Biotecnologia, Universidade Católica Portuguesa, Rua de Diogo Botelho 1327, 4169-005 Porto, Portugal
* Correspondence: slawomir.ciesielski@uwm.edu.pl (S.C.); ivmoreira@ucp.pt (I.V.-M.)

Citation: Ciesielski, S.; Vaz-Moreira, I. Advances in Industrial and Environmental Microbiology. *Appl. Sci.* **2021**, *11*, 9774. https://doi.org/10.3390/app11209774

Received: 30 September 2021
Accepted: 14 October 2021
Published: 19 October 2021

1. Introduction

Understanding microorganisms in terms of their functionality, diversity, and interactions with other organisms is crucial for better understanding of our Biosphere. However, this can be difficult, not only because of their small size, but also because of their rapid evolution and tendency to create multispecies communities. The structures of microbial communities are closely associated with environmental conditions and therefore are likely to evolve in the context of global change [1]. Microbial communities can be divided into subcommunities, based on different attributes, such as abundance, functional type, and activity. They can differ in their environmental sensitivity, interaction, and distribution patterns [2].

The main impediment to a better understanding of the structure and function of microbial communities has been a lack of research tools that are sensitive enough. For decades, moreover, researchers have tried to cultivate microorganisms in laboratory conditions with only limited success. The rapid evolution of these organisms causes their physiological requirements to change faster than was expected, and growth conditions cannot be established for many of them. Therefore, methods based on analysis of nucleic acids recovered directly from environmental samples have become standard approaches for the characterization of microbial communities. In particular, next-generation DNA sequencing methods and in silico analyses have allowed researchers to address problems associated with unculturable microorganisms [3]. The metaomics approaches, including phylogenetic marker-based microbiome profiling, shotgun metagenomics, metatranscriptomics, metaproteomics, and metabolomics, have not only allowed identification of unculturable microorganisms but have also provided information about their physiological status.

Due to their ubiquity, microorganisms have developed innumerable substrate utilization mechanisms, which allow them to survive and evolve in even the most difficult environmental conditions. Their metabolic versatility means that microorganisms could be the source of various biochemicals, such as drugs, biopolymers, and enzymes. Additionally, this metabolic versatility makes microorganisms invaluable for environmental protection, especially in bioremediation, wastewater treatment, and conversion of biomass into energy.

This Special Issue on "Advances in Industrial and Environmental Microbiology" presents a collection of articles that focus on both applied and environmental microbiology, which are briefly summarized here for the convenience of the reader.

2. Industrial Microbiology

The potential and recent applications of xanthan gum biopolymer, with a particular focus on improving soil properties is reviewed by Mendonça et al. [4]. Xanthan gum is a polysaccharide formed by aerobic fermentation of sugar by *Xanthomonas campestris*. It is commonly used as a food additive because it can serve as a hydrocolloid rheology modifier. Soil treatment with xanthan gum has been shown to induce partial filling of soil voids and the generation of additional links between soil particles, which decreases the permeability coefficient of the soil and changes its mechanical properties. Via the bioclogging method, xanthan gum can be used to enable the filling of soil pores and the formation of a barrier that can increase the strength and cohesion and reduce the permeability coefficient of the soil. The authors conclude that these findings should be tested in practice to support the results obtained from laboratory experiments.

The treatment of wastewater in sequencing batch reactors with a high concentration of suspended solids from meat processing is the subject of a research article authored by Jachimowicz et al. [5]. The main aim of their study was to explore interrelationships between the aeration mode, microbial community composition, and the efficiency of removal of nitrogen, phosphorus, organic compounds, and total suspended solids. The authors show that intermittent aeration significantly increases denitrification and phosphorus removal efficiencies but leads to the decomposition of extracellular polymeric substances. In addition, they show that the microbial structure changes significantly over time. They report that, in mature granules during the period of stable treatment of meat-processing wastewater, an increased relative abundance of *Arenimonas* sp., *Thauera* sp., and *Dokdonella* sp. is typical. Additionally, they show that constant aeration favors the growth of microorganisms from the *Rhodanobacteraceae*, *Rhodobacteraceae*, and *Xanthomonadaceae* families, while intermittent aeration favors the growth of the *Competibacteraceae* family.

3. Environmental Microbiology

Recent advances in understanding of the role of microbial communities in subterranean ecosystems, namely their contribution to biogeochemical processes affecting the composition of the subterranean atmosphere and the availability of nutrients is reviewed by Martin-Pozas et al. [6]. These bacterial communities have been found to actively contribute to the uptake–fixation-production of CO_2, CH_4, and NxOx gases. In addition, the bacteria from these still poorly explored niches display potential to serve as producers of bioactive compounds with antibacterial, antifungal, or anticancer activities, as well as enzymes with biotechnological applications. Moreover, methanotrophs and heterotrophs in subterranean environments can interact and stimulate each other's growth, and their growth is associated with the production of bioactive compounds. The advent of massive sequencing technologies and the omics sciences have brought to bear important tools for continued exploration of the potential of biological processes in subterranean ecosystems.

Taking into consideration the origin and proportion of wastewater effluents (urban, hospital, livestock farms, or industrial), Rolbiecki et al. [7] do an analysis of the efficiency of wastewater treatment plants (WWTPs) in reducing the prevalence of antibiotic resistance genes (ARGs). They also examine the effect of the treated wastewater on the receiving water bodies (rivers). Their first conclusion is that the prevalence of ARGs in raw wastewater is affected by the origin of the sewage. For example, agricultural sewage containing animal waste is associated with a higher abundance of the *sul2* gene, while hospital wastewater appears to be the main source of the *sul1* gene. The authors also conclude that WWTPs are important sources of dissemination of ARGs to receiving water bodies. Metagenomic and metatranscriptomic analyses will be applied to these samples to increase knowledge about system dynamics, to develop effective strategies for managing risks, and to evaluate effects on human health.

In some natural ecosystems, invasion by exotic plants poses a threat to the equilibrium of the environment. For this reason, Cao et al. [8] studied the influence of Alternanthera philoxeroide (alligator weed) invasion on wetland organic matter accumulation and bacterial community changes. Samples were collected from two sediment depths (0–15 and 15–25 cm) in invaded and normal habitats in two natural wetlands and two constructed wetlands. They observed that the contents of the light fractions of organic carbon and nitrogen were significantly higher in the constructed wetlands than in the natural wetlands, and the contents of the C and N fractions were also higher in the invaded areas than in the normal wetland habitats. The presence of the invasive weed also correlated with changes in the wetlands' microbiota. Although *Proteobacteria*, *Bacteroidetes*, *Acidobacteria*, and *Chloroflexi* were the dominant bacterial phyla in the wetlands, the abundance of *Acidobacteria*, *Actinobacteria*, and *Gemmatimonadetes* was significantly higher in invaded areas.

4. Conclusions

The scientific articles covered by this Special Issue show that the study of bacterial communities as a whole is essential for understanding the dynamics of both industrial and natural environments. The interactions between the different members of a bacterial community determine not only the structure but also the composition of the community. Additionally, the presented papers, although they approach different problems, show that rapid progress is possible thanks to modern research tools. In particular, the use of omics tools has become an important part of all fields of microbiological study.

Author Contributions: Conceptualization, S.C. and I.V.-M.; writing—original draft preparation, S.C. and I.V.-M.; writing—review and editing, S.C. and I.V.-M. All authors have read and agreed to the published version of the manuscript.

Funding: This research received no external funding.

Acknowledgments: I.V.-M. would like to thank the scientific collaboration under the FCT project UIDB/50016/2020.

Conflicts of Interest: The authors declare no conflict of interest.

References

1. Gutknecht, J.L.M.; Field, C.B.; Balser, T.C. Microbial communities and their responses to simulated global change fluctuate greatly over multiple years. *Glob. Chang. Biol.* **2012**, *18*, 2256–2269. [CrossRef]
2. Liu, J.; Meng, Z.; Liu, X.; Zhang, X.-H. Microbial assembly, interaction, functioning, activity and diversification: A review derived from community compositional data. *Mar. Life Sci. Technol.* **2019**, *1*, 112–128. [CrossRef]
3. Malla, M.A.; Dubey, A.; Yadav, S.; Kumar, A.; Hashem, A.; Abd_Allah, E.F. Understanding and designing the strategies for the microbe-mediated remediation of environmental contaminants using omics approaches. *Front. Microbiol.* **2018**, *9*, 1132. [CrossRef] [PubMed]
4. Mendonça, A.; Morais, P.V.; Pires, A.C.; Chung, A.P.; Oliveira, P.V. A Review on the Importance of Microbial Biopolymers Such as Xanthan Gum to Improve Soil Properties. *Appl. Sci.* **2021**, *11*, 170. [CrossRef]
5. Jachimowicz, P.; Cydzik-Kwiatkowska, A.; Szklarz, P. Effect of Aeration Mode on Microbial Structure and Efficiency of Treatment of TSS-Rich Wastewater from Meat Processing. *Appl. Sci.* **2020**, *10*, 7414. [CrossRef]
6. Martin-Pozas, T.; Gonzalez-Pimentel, J.L.; Jurado, V.; Cuezva, S.; Dominguez-Moñino, I.; Fernandez-Cortes, A.; Cañaveras, J.C.; Sanchez-Moral, S.; Saiz-Jimenez, C. Microbial Activity in Subterranean Ecosystems: Recent Advances. *Appl. Sci.* **2020**, *10*, 8130. [CrossRef]
7. Rolbiecki, D.; Harnisz, M.; Korzeniewska, E.; Jałowiecki, Ł.; Płaza, G. Occurrence of fluoroquinolones and sulfonamides resistance genes in wastewater and sludge at different stages of wastewater treatment: A preliminary case study. *Appl. Sci.* **2020**, *10*, 5816. [CrossRef]
8. Cao, Q.; Zhang, H.; Ma, W.; Wang, R.; Liu, J. Composition characteristics of organic matter and bacterial communities under the Alternanthera philoxeroide invasion in wetlands. *Appl. Sci.* **2020**, *10*, 5571. [CrossRef]

MDPI

Review

A Review on the Importance of Microbial Biopolymers Such as Xanthan Gum to Improve Soil Properties

Amanda Mendonça [1], Paula V. Morais [2,*], Ana Cecília Pires [2], Ana Paula Chung [2] and Paulo Venda Oliveira [3,*]

[1] Department of Mechanical Engineering, University of Coimbra, 3030-788 Coimbra, Portugal; amaandamendonca@hotmail.com
[2] CEMMPRE, Department of Life Sciences, University of Coimbra, 3000-456 Coimbra, Portugal; accp@ua.pt (A.C.P.); apchung@gmail.com (A.P.C.)
[3] ISISE, Department of Civil Engineering, University of Coimbra, 3030-788 Coimbra, Portugal
* Correspondence: pvmorais@ci.uc.pt (P.V.M.); pjvo@dec.uc.pt (P.V.O.)

Abstract: Chemical stabilization of soils is one of the most used techniques to improve the properties of weak soils in order to allow their use in geotechnical works. Although several binders can be used for this purpose, Portland cement is still the most used binder (alone or combined with others) to stabilize soils. However, the use of Portland cement is associated with many environmental problems, so microbiological-based approaches have been explored to replace conventional methods of soil stabilization as sustainable alternatives. Thus, the use of biopolymers, produced by microorganisms, has emerged as a technical alternative for soil improvement, mainly due to soil pore-filling, which is called the bioclogging method. Many studies have been carried out in the last few years to investigate the suitability and efficiency of the soil–biopolymer interaction and consequent properties relevant to geotechnical engineering. This paper reviews some of the recent applications of the xanthan gum biopolymer to evaluate its viability and potential to improve soil properties. In fact, recent results have shown that the use of xanthan gum in soil treatment induces the partial filling of the soil voids and the generation of additional links between the soil particles, which decreases the permeability coefficient and increases the mechanical properties of the soil. Moreover, the biopolymer's economic viability was also analyzed in comparison to cement, and studies have demonstrated that xanthan gum has a strong potential, both from a technical and economical point of view, to be applied as a soil treatment.

Keywords: geotechnical engineering; microbiology; biopolymer; bioclogging; xanthan gum; soil improvement

Citation: Mendonça, A.; Morais, P.V.; Pires, A.C.; Chung, A.P.; Oliveira, P.V. A Review on the Importance of Microbial Biopolymers Such as Xanthan Gum to Improve Soil Properties. *Appl. Sci.* **2021**, *11*, 170. https://dx.doi.org/10.3390/app11010170

Received: 27 November 2020
Accepted: 23 December 2020
Published: 27 December 2020

1. Introduction

The construction of the current civil engineering structures, such as buildings and embankments, is only viable if the foundation soil demonstrates suitable geotechnical characteristics. In the presence of soils with poor geotechnical properties, Geotechnical Engineering uses several techniques (such as shallow and deep compaction, stone and concrete columns, the preloading technique, staged construction, shallow and deep soil stabilization with cement or other binders) to improve the soil's properties, such as strength, stiffness, erosion resistance, shear behaviour, plasticity, porosity and hydraulic conductivity, among others [1]. Since the beginning of human civilization, several different methods and materials have been used to improve the properties of soils. Especially after the Industrial Revolution, Portland cement became the most commonly used material for construction and soil treatment [2,3].

Portland cement is a material that provides many well-known benefits in the construction industry, such as the low cost, durability, high strength, and workability of the material. However, the process of production and application of the cement is linked to some environmental concerns. The CO_2 emissions related to the use of cement are reported to be one of the world's leading causes of CO_2 emissions, and represent a percentage of almost 8% of all global CO_2 emissions [1,3]. In order to reduce these environmental impacts, the scientific community has been developing new, more suitable applications with reduced pollution emission [4]. Materials that are described as environmentally friendly have been developed for application using sustainable methods, namely microbial biopolymers (i.e., high molecular polysaccharides produced by microbes). These biopolymers have been introduced as an alternative to conventional techniques for soil treatment [5,6]. Several studies have shown a great strengthening of soil after the application of biopolymers due to direct ionic bonding with fine particles or to the formation of a continuous biopolymer matrix between coarse particles. Therefore, Geotechnical Engineering has identified various potential applications for these biopolymers to solve soil problems [2]. One of the promising applications considered is bioclogging, which promotes the production of pore-filling materials in the soils through microbial means, and thus reduces the porosity and hydraulic conductivity of the soil [6]. In recent years, several studies have been published on the applicability of biopolymers, including experimental results and their analyses, but until now, there have been few cases of practical implementation [4–6].

2. Biological Processes in Geotechnical Engineering

Over the years, most engineers and researchers have not given adequate attention to biological processes in the soil, but the demand to satisfy society's concerns about environmental issues has forced researchers to focus on new procedures that combine the use of microorganisms in innovative soil improvement techniques [7]. As a result, the amount of research involving microbial activities to improve the soil properties required by engineering has increased and, simultaneously, so has the intention to lower rates of pollutant gas emissions when compared to the usual geotechnical practices that use Portland cement [5].

The part of Geotechnical Engineering that uses biological processes to solve geotechnical difficulties is called Biogeotechnology [2]. In other words, this is a field related to the chemical reactions achieved and measured within the soil by means of biological action [2,3]. Important factors such as the selection and proof of appropriate microorganisms for specific applications have to be considered, as do the cost-effectiveness and biosafety of the application [5]. Several studies on biotechnological engineering applications already exist, such as the use of vegetation, algae, bacteria, enzymes, and biopolymers. However, two of them, namely biocementation and bioclogging, have received special attention due to the favorable results presented by applications that modify soil's properties, in a way that is favorable to engineering, by increasing stiffness and strength and even reducing permeability [5,8]. Biocimentation, usually based on calcium carbonate precipitation, can be promoted by microorganisms (MICP—microbial induced calcium carbonate precipitation) or by enzymes extracted from bacteria or plants (EICP—enzyme-induced calcium carbonate precipitation). These two methodologies rely on the same chemical reactions that decompose urea into carbonate and ammonium ions. The carbonate ions combine with calcium ions to form precipitated calcium carbonate ($CaCO_3$) [9]. Bioclogging relies on the use of biopolymers. These bio-mediated soil improvement techniques are low-cost and require minimal extra energy, which guarantees a reduction in the carbon dioxide released and thus also contributes to reducing greenhouse gases [10].

2.1. Application of Biopolymers in Soil Treatment

Three main typical types of main biopolymers are considered: polynucleotides (for example, RNA and DNA), polypeptides (amino acid compounds), and polysaccharides, which have been the type most commonly applied in various engineering practices [6,8,10]. However, the use of biopolymers is not, in fact, an entirely new method in geotechnical engineering, since in a broad sense, organic polymers such as natural bitumen, straw, and rice have already been used in ancient civilizations and can also be classified as biopolymers.

Studies have shown that the direct use of biopolymers in soil can have several advantages over traditional biological treatment methods. This fact can be explained because different types of biopolymers have shown a good interaction with the soil. Additionally, biopolymers have distinct properties such as good viscosity, resistance to shear degradation, and stability over wide pH and temperature ranges, which can promote significant strengthening effects in soils by enhancing cohesion, strength and resistance to erosion, and reducing its permeability [10].

The effect of biopolymers on the soil is related to their ability to make a stable gel matrix within the soil without damaging the local ecosystem. Moreover, they have the potential to promote vegetation growth [11], which is also beneficial in terms of stabilizing shallow soil, since it contributes to increasing soil resistance to erosion and to the satility of slope. The industrial production of biopolymers of microbial origin, such as chitosan and xanthan gum, creates biopolymers with better workability in terms of material rheology. Table 1 summarizes some characteristics of the common biopolymers used in geotechnical engineering [6,8,12].

Table 1. Effect of common biopolymers on soil properties.

Biopolymer	Effect
Casein and Sodium Caseinate salt	Stabilize sand dunes [1].
Curdlan	Clogs the soil [2].
Chitosan gum	Decreases hydraulic conductivity [11]; Remediates wastewater and contaminated soil [1,5].
Xanthan gum	Decreases permeability; Retains water due to strong hydrogen bonding [1,12,13].
β-Glucan	Increases the resistance to soil erosion and promotes the growth of vegetation [1,14].
Gellan Gum	Improves mechanical properties [1,4,14,15].
Polyacrylamide (PAM)	Decreases soil erosion due to PAM hydrogels [1,15].
Agar Gum	Rapid gelation and increases water infiltration [16].
Guar Gum	Increases mechanical properties [1,4,17–19]; Reduces the collapsible potential of plastic soil [13]; Decreases the hydraulic conductivity of silty sand [20]; Increases the liquid limit of kaolinite clay [21].

2.2. Soil–Biopolymer Interaction

Recent studies indicate that the effect of the use of biopolymers in soil treatment is due to the association and the balance of two main factors, the creation of additional links between the soil particles and the filling of the soil's voids (at least partially) with hydrogels. The creation of links in the soil skeleton explains the better effectiveness of this methodology in strengthening clayey soils in relation to sandy soils [1]. In fact, biopolymers have high specific surfaces with electrical charges that enable them to interact directly with the clay's minerals, and also provide firm matrices of high strength biopolymer-soil due to the electrical charges [6]. This type of link is responsible for the increase in the unconfined compressive strength and cohesion resistance observed in clay–biopolymer mixtures in relation to the results obtained with pure sands [1,12].

Although the effect of biostabilisation with biopolymers is higher in the presence of clays than in sands, the experimental results show that the impact of this methodology in sandy soils is also positive. The SEM (Scanning Electron Microscopy) image of the structure of a sand stabilized with xanthan gum (Figure 1) clearly shows that the stabilization of sandy soils with biopolymers creates a network of links between the soil particles.

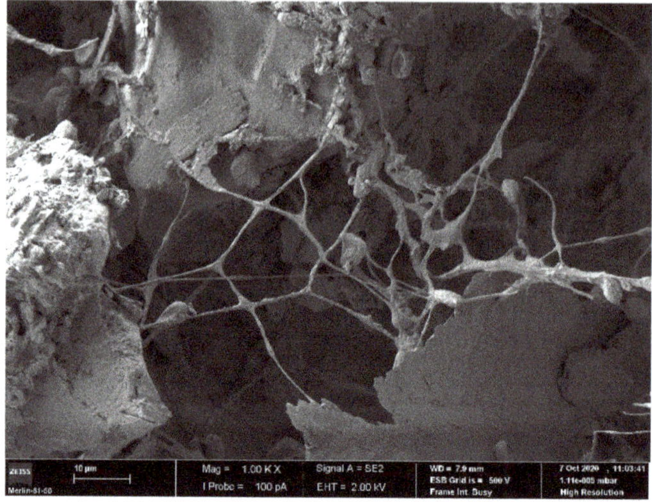

Figure 1. SEM image of the structure of a sandy soil stabilized with xanthan gum [22]. Previously, a thin layer of gold coating was deposited on the sample by sputtering. SEM tests were performed in accordance with Pansu and Gautheyrou [23].

Previous studies have demonstrated that this methodology is suitable for use in the majority of natural soils, which are composed of a mixture of coarse and fine particles (silt and clay), where there is a combination of the strength of the biopolymer–clayey-soil matrices (improvement in cohesion), and the aggregation of coarse particles (improvement in the friction angle) can be observed [11].

Taking these results into consideration, a previous in situ soil characterization is recommended to better understand its composition and to determine the most appropriate method and biopolymer to maximize the strengthening effect [12].

A second effect of the use of biopolymers in soil treatment is the formation of viscous hydrogels capable of filling the soils' pores spaces, especially in sand, which causes the bioclogging effect, which has, as a main consequence, the reducing of the hydraulic conductivity by about 3–4 orders of magnitude [13]. Therefore, biopolymers have the potential to be applied in several hydraulic purposes such as grouting, slurry walls and infiltration barriers (temporary) [14].

The amount of water is another essential factor to be considered in soil–biopolymer interaction. A study by Chang et al. [1] showed that when soils have a higher water content, the system formed by the soil and the biopolymer, whether it is a dehydrated gel or a matrix, has the ability to adsorb and transfer water to hydrogels. This induces the volumetric expansion (swelling) of the soil [11,13] and a consequent increase in the distance between soil particles, leading to an increase in the soil-void ratio of the soil [24].

In fact, a recent study shows that the void ratio fundamentally increases during the first 7 days of curing time, which corresponds to the time required to hydrate the hydrogels; after that, a slight decrease in the void ratio is observed due to some dehydration of the hydrogels [24]. Therefore, some soil properties, specifically its tensile strength and stiffness, proportionally decrease with the increase in the water content and can induce a remarkable reduction in soil strength, approximately 10 times less than that which occurs in a fully saturated condition [1]. The ability of the soil–biopolymer mixtures to retain water is also confirmed by the increase in the liquid limit (LL) of the biostabilized soil with the increase in the biopolymer content [25].

3. Bioclogging Technique

The bioclogging technique is an aspect of Microbial Geotechnology used to reduce the permeability of soils and some porous rocks as a result of the accumulation of bacterial biomass and the growth of biofilms on the surface they are applied to [10].

One of the main results of the bioclogging process is the microbial production of a water-insoluble biopolymer [15]. However, to effectively adopt this microbial methodology, several factors must be considered, such as the screening and identification of suitable microorganisms to the different applications and environments. Besides the optimization of microbial activity, its biosafety, the cost–benefit ratio and the stability of the soil's properties after interaction with the biopolymer (i.e., biomodification) [2] must be evaluated. Engineers consider that bioclogging could be applied to solve engineering problems related to soils such as by diminishing the hydraulic conductivity of earthworks, reducing infiltration of the ponds and leakage in construction sites, preventing channel erosion and forming grout curtains to lower the migration of heavy metals or organic pollutants [11,12].

3.1. Bioclogging as Solution in Geotechnical Engineering

The integration of engineering, microbiological and ecological studies is required to implement biological methods to improve the mechanical properties of soils, and bioclogging has an important role in this solution. Bioclogging involves the use of microorganisms, especially the ones that produce polysaccharides, which results in a positive correlation between the microbial biomass produced and the stability of soil [5,15].

Since the 1990s, some studies have attempted to use bioclogging tests to reduce soil permeability and diminish soil erosion [9]. Many water-insoluble polymers and gel formers of bacterial origin are produced by industry; these materials still only have limited use due to the cost involved and limited knowledge [11,14].

Consequently, laboratory tests have been carried out to identify the microbial groups that can be used for bioclogging. Ivanov et al. [2] have shown that cultures of nitrifying and oligotrophic bacteria can be used for this purpose. The culture of autotrophic nitrifying bacteria grown in sand with dissolved ammonium produces microbial polysaccharides from the ammonium and from CO_2 in the air, which can reduce the hydraulic conductivity of the sand by about 10^{-4} m/s to 10^{-6} m/s. In the case of oligotrophic heterotrophic bacteria, they were grown in the sand with a low glucose concentration and produced polysaccharides that were also capable of reducing hydraulic conductivity in the same order [2,8]. Some of the main different applications of bioclogging are presented in Table 2, showing their wide range, as well as some of the mechanisms and essential conditions for effective bioclogging [16,17].

Table 2. Potential applications of bioclogging in Geotechnical Engineering (based on Ivanov et al. [2]).

Group of Microorganisms	Bioclogging's Mechanism	Essential Conditions	Potential Geotechnical Applications
Algae and cyanobacteria	Formation of impermeable biomass layer	Penetration of light and presence of nutrients	Reduction in water infiltration
Aerobic and facultative anaerobic slime-production bacteria	Production of slime in soil	Presence of oxygen and medium with ratio of C:N > 20	Avoidance of cover for soil erosion control and slope protection
Nitrifying bacteria	Production of slime in soil	Presence of ammonium and oxygen in soil	Reduction in drain channel erosion
Sulphate-reducing bacteria	Production of undissolved sulfides of metals	Anaerobic conditions, presence of sulfate and a source of carbon in the soil	Formation of grout curtains to reduce the migration of heavy metals.
Ammonifying bacteria	Formation of undissolved metals carbonates in the soil due to an increase in pH and release of CO_2	Presence of urea and dissolved metal salt	Prevention of piping of earth dams and dikes.

A study by Kucharski et al. [26] also claims that bioclogging can be used in industry to reduce fluid flow, improve oil recovery from reservoirs and repair cracks in concrete [16]. Bioclogging, as well as biocementation, is a technique that has several features and has the potential to grow even more due to all the advantages presented in its use. What cannot be forgotten is the fact that it is a technique directly linked to the type of biopolymer used, so it is always necessary to choose the most appropriate type for the expected target [18].

3.2. *Bioclogging Application: Benefits and Limitations*

The bioclogging method has several advantages in the field of engineering when compared to other biological soil treatment methods. Biopolymer can have a direct application since it can be produced both in-situ and ex-situ (i.e., exo-cultivation). When biopolymers are produced ex-situ, both quality and quantity can be controlled to ensure a proper bioclogging process. Apart from this, biopolymers can be produced in large quantities, react with soil particles immediately and can be applied in many different ways such as by mixing, injection or spraying, which allows their application for both temporary or rapid supporting purposes [9,14]. However, the greatest advantage of the bioclogging method is that it can effectively replace the more energy-demanding mechanical compaction and also the more expensive or environmentally unfriendly chemical grouting methods [15]. Bioclogging methodology also reduces carbon gas emissions, supports vegetation growth and promotes the soil stabilization, which can contribute to the farmland preservation. It also has anti-desertification effects and may be used in response to other threats to environmental conservation [16].

On the other hand, the most concerning factor of the bioclogging technique is the concentration of dissolved oxygen and destabilization of soil's properties after treatment [12]. Furthermore, the concentration of carbon, nitrogen, and metabolites such as hydrocarbons or organic acids will also largely affect the growth and activity of the clogging microorganisms in the soil. The biopolymers applied to soils are organic materials, so there must be a concern with their biodegradability over time. They are also highly sensitive to the presence of water and are subjected to cycles of wetting and drying, which can affect their durability in soil, and consequently, the strength of soils in the presence of water [18,19].

Another limitation is related to the depth reached by microbial clogging in situ since the penetration of microbial cells generally only occurs to a limited depth. Therefore, the bioclogging technique can only be applied to a limited type of soils, normally sand or clay with adequate hydraulic conductivity and several other challenges are related to this use. However, the wide variety of biopolymers available and their flexibility makes bioclogging an effective option [20].

Biopolymers that are used in bioclogging have some important characteristics, such as high viscosifying power, high resistance to shear degradation, pseudoplasticity, and stability at various ranges of temperature and pH. They are used to fill pores in granular media and, therefore, to reduce hydraulic conductivity and strengthen the material through cementation [15,20]. Characteristically, xanthan gum, cellulose, starch, gel gum, guar gum, agar gum and curdlan are biopolymers that show gelation properties, granting them the potential to be used in bioclogging techniques [1]. The high pseudoplasticity and the ability to increase the viscosity of the medium when mixed with it makes xanthan gum one of the best candidates to be used in soil bioclogging applications.

4. Xanthan Gum Biopolymer

4.1. *Xanthomonas Campestris Species*

Xanthan gum ($C_{35}H_{49}O_{29}$) is a polysaccharide formed by aerobic fermentation of sugar of the bacterial species *Xanthomonas campestris* and is commonly used as a food additive due to its characteristics as a hydrocolloid rheology modifier [20,21]. It is an anionic polysaccharide composed of D-uronic acid, D-mannose, pyruvylated mannose, 6-0-acetyl D-mannose, and 1,4-linked glucan [19]. The chemical structure is led by the linear chain 1,4-linked β-D glucose backbone with two tri-saccharide elements pairing the side chain on every glucose element. The other side chain is composed by a D-glucuronic acid element-linked between two D-mannose elements [15,19]. The xanthan gum-produced by the type strain of the bacterial species is considered by the US Food and Drugs Administration (FDA) as an additive for industrial food; it is of great interest because it has rheological characteristics not found in other polymers [16]. *Xanthomonas campestris* is an aerobic and microaerophilic bacterium, so it is easily cultivated in the laboratory with an optimum growth temperature of about 25–30 °C. This species grows rapidly and produces turbidity in the medium within approximately 2 or 3 days of cultivation with an optimal pH between 6 and 8 [17].

4.2. General Features

High stability under a wide range of temperature and pH, and also the pseudo-plasticity (i.e., viscosity degradation), are important factors that make xanthan gum suitable for diverse applications [12]. This biopolymer can be mixed with both cold and hot water, but it is still highly viscous due to its ability to form a viscous hydrogel when mixed with water. This viscous hydrogel is produced through the absorption of water molecules by the hydrogen bonds that compose xanthan gum [27,28].

Chang et al. [17] showed that a proportion of almost 0.5% of xanthan gum in relation to the weight of the soil is able to cause almost the same effect on the resistance to soil erosion when the equivalent of 10% of cement is used. In the specific case of Korean red-yellow soil–xanthan gum treatment, it also improved the growth of vegetation, a condition explained by its strong water adsorption during the rainy seasons and also high soil moisture retention during the dry season [29]. However, while xanthan gum or other biopolymers in the polysaccharide group have been shown to be effective even in small concentrations, what may still be a major limitation concerning its use is the issue of the price [13]. Therefore, Chang et al. [1] made a specific analysis that can prove that it is no longer valid to affirm that the cost of xanthan gum can be considered a problem. Instead, this review continued the analysis from 2015 up to 2020 and showed that, due to the increase in its use and production in large quantities, the cost of xanthan gum has reduced significantly over the last thirty years (Figure 2), that is, the cost of adding 0.5% of xanthan gum for soil treatment (i.e., 5 kg of xanthan gum per 1 ton of soil) has decreased from approximately 70 USD to 10 USD per ton of soil over the last three decades.

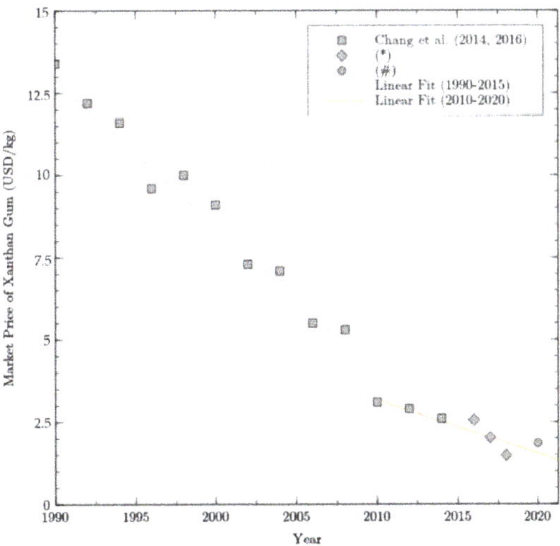

(*) https://www.adroitmarketresearch.com/industry-reports/xanthan-gum-market
(#) https://www.echemi.com/productsInformation/pd20150901009-xanthn-gum.html

Figure 2. Decrease in the market price of the xanthan gum during the last 30 years.

Although the use of most biopolymers currently represents a high cost for Geotechnical Engineering, further research on the subject is necessary [12]. Additionally, most of the market prices presented are intended for use in the food industry, which requires very high purity and, consequently, results in a significantly higher production cost. For geotechnical applications, such levels of purity are unnecessary and, therefore, the price of biopolymers can reasonably be expected to decrease when they are produced specifically for this purpose [12,25].

Figure 3 shows the data obtained in Jang's research [28] that compared the price of the five most common biopolymers used for soil stabilization, and it may be seen that xanthan gum and the non-biological toxic polymer polyacrylamide (PAM) cost less than 1700€/ton, which is considerably cheaper than the other three polymers. For this reason, xanthan gum may definitely be considered a good option for geotechnical purposes, given the economic aspect that has only improved over the years and, in some countries, it may even represent a cheaper option than ordinary cement [12].

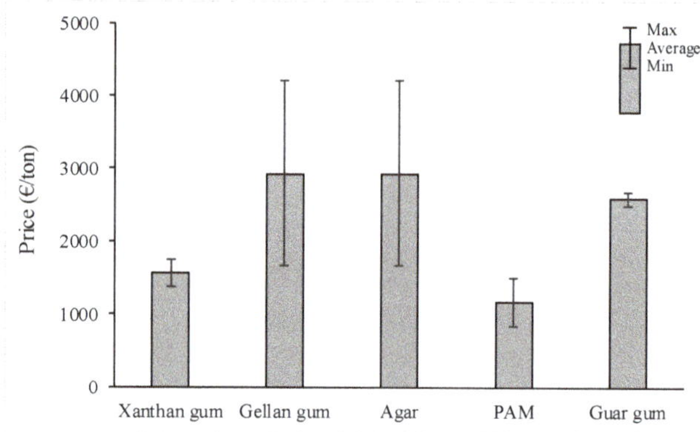

Figure 3. Prices of five common biopolymers (based on Jang [28]).

4.3. *Effects of Xanthan Gum on Soil Properties for Geotechnical Engineering*

Recent studies have shown the good performance of xanthan gum used in soils [12]. The biopolymer content, soil type and density, hydration level, mixing method and void ratio are considered the major parameters that control the use of biopolymer for shear strength and reduction in hydraulic conductivity in the field of soil improvement. Therefore, especially considering the use of xanthan gum, it also has other influencing factors, such as the presence of ions (i.e., alkali or alkali–earth metal ions) that enable the interaction (i.e., hydrogen bonding) between the inter-particle and the charged surfaces of soil that allow the formation of matrices and are similar to a hard plastic. It is expected to provide stronger hydrogen or electrostatic bonding governed by the strength of the xanthan gum between their monomers and soil particles. In addition, this biopolymer rheology increases the benefits of its use, and it has been observed that xanthan gum has better effect and workability with well-graded soils with fine particles [15,29].

4.3.1. Mechanical Properties

A study from Lee et al. [30] concerned the strength behaviour of a sand–xanthan gum mix in three different phases of the gel: initial, dry and re-submerged. For the initial phase, the test was performed as soon as the samples were prepared; in the dried phase, the samples were in a drying room for 28 days; and the re-submerged samples were submerged in distilled water for 24 h before performing the tests. The study concluded that even a very small addition of xanthan gum to the sand can change the strength behaviour of this type of soil. The tests proved that increasing the xanthan gum content in the initial gel increases cohesion, but the friction angle remains constant. In the dry gel phase, both the cohesion angle and friction increased with the addition of more xanthan gum content due to the formation of the biofilm that occurs on the surface of particles, as well as the formation of viscous hydrogels that induces the bioclogging effect. Otherwise, the friction angles of the sand–xanthan gum mixture decreases as the xanthan gum content increases (above 1%) in the submerged condition. This is because the interaction on the surface friction between inter-particles is reduced due to the increased pressure generated [11,13,30]. One of the reasons that explain the improvement in the shear strength in soil–xanthan gum mixtures is due to the presence of different chemical functional groups, such as hydroxyls and esters. Moreover, its long chain also allows additional locations in which the characteristic chemical reaction of a specific functional group can occur. As described for other polymers, the chemical bonds between the biopolymer and the soil particles are formed due to the adhesive forces that hold the soil particles on their surfaces. At microscopic scales, this effectiveness is observed through connective forces existing at the gel–particle interface, which are forces that contain and allow ionic/electrostatic or covalent bonds (i.e., chemisorption), hydrogen bonds (i.e., strong polar attraction) and Van der Waals bonds (i.e., physical absorption) [19,22,24,28].

Tests performed to determine the effect of xanthan gum content (from 0.25% to 2%) on the modulus of elasticity of biostabilized sandy and silty soils by Ayeldeen et al. [29] show, for both soils, that the modulus of elasticity increases with the biopolymer content.

Several studies have shown that xanthan gum can be used successfully to improve the unconfined compressive strength and stiffness of sands [17,31–33], silty sands [34], clays [17,35], residual soils [36], bentonite, kaolinite [17,25,37], and soft marine clays [24], as well the undrained shear strength of mine tailing material [38]. Additionally, experimental results revealed that the most favourable xanthan gum content for soil stabilization is in the range of 1–1.5% for ambient curing and oven curing [39] and 2.5% for curing in submerged conditions [35]. Soldo et al. [40] concluded that the optimum xanthan gum content depends on the soil type and water content.

Experimental results also showed that the mechanical properties (unconfined compressive strength and stiffness) tend to improve with curing time (up to 28 days) for bentonite and kaolinite [39,40], clays [17,35] and sandy soils [17,31,32,40]. A similar time evolution was obtained for the shear strength and its strength parameters (cohesion and angle of shear strength) of clayed soils obtained from direct shear tests [35,39]. Although this tendency was confirmed for a xanthan gum-treated bentonite and kaolinite for a curing time up to 90 days [39], the results obtained from a biostabilized sand with silt indicate that the soil strength does not change considerably after a curing time of 5 days [40]. These differences in the behaviour are probably due to the fact that the bonds promoted by the xanthan gum are less exposed to the environment in a clayed soil than in pure sand, where a faster interaction of xanthan gum with water occurs [40].

As previously mentioned, the increase in the mechanical properties over time is related to dehydration of the hydrogels associated with the formation of a soil–xanthan gum matrix, which promotes the aggregation of soil particles [17,34]. Besides the positive effect on the stiffness, compressive and shear strength when using xanthan gum to stabilize soils, the results of direct tensile strength tests [41] and splitting tensile strength tests [40] carried out with air-dried specimens of two sandy soils show an increase in the tensile strength with the increase in the xanthan gum content (from 1.5 to 3%) and with the curing time (up to 30 days). This effect is justified by the combination of higher suction and hydrogel bonding [41].

Recent research about the behaviour of xanthan gum biopolymer and sand interaction during drying indicates that this biopolymer promotes a reduced effect on shear strength when the water content is high, independently of the biopolymer content [42]. On the other hand, the continuous evaporation of the water induces a significant increase in the shear strength of the xanthan gum treated soil and this effect is higher with the increase in the xanthan gum content. Additionally, the strengthening is more significant when the drying was carried out in an oven at 40 °C than at a room temperature of 20 °C. In fact, the latter drying conditions bring about a reduced increase in shear strength, since under these conditions the drying and the crystallization (promoted by the xanthan gum) is fundamentally superficial, while, in the inner part of the sample, the water content remains high enough to only generate weak bonds [42].

4.3.2. Permeability

In terms of the permeability effect, experiments by Ayeldeen et al. [29] clearly showed that xanthan gum has a remarkable effect in reducing soil permeability, in both sand and silt mixture. After 5 weeks of curing time, sand samples treated with 0.25% xanthan gum had a permeability reduction of almost 60% compared to the initial value. The same trend was observed for mixtures with silt; however, the relative amount of reduction in permeability was slightly lower in mixtures with sand.

According to Czarnes et al. [43], the reduction in permeability is a result of the cross-linking elements that form within the soil matrices and fill in the voids that can obstruct the flow of water. Another relevant factor emphasized in research studies is the curing time effect, especially related to permeability tests of xanthan gum-treated soil mixtures. Bouazza et al. [44], Dehghan et al. [45] and Cabalar et al. [31] have also stated that the permeability increases proportionally to the curing time, but this increase rate can change according to the soil type and xanthan gum concentration. This permeability rate increase is due to dehydration that causes shrinkage of the biopolymer matrices inside the gaps in the soil [25,26]. In tests by Ayeldeen et al. [29] with a sand–xanthan gum treated mixture, the permeability increased in the range of 150 to 220% between the first and the tenth week; however, for the silt–xanthan gum mixture, it was between 115 and 145% in the same period of time. For this case, the researchers state that the permeability increases more in the sand mixture instead of the silt mixture because the fine particles present in this type of soil delay the dehydration process and, consequently, the evaporation of water in the empty spaces of the soil.

Mendonça [22] compared the effectiveness of the commercial xanthan gum with a xanthan-like gum obtained directly from a strain of *Stenotrophomonas* sp. on the permeability of sandy soil. The results from both types of xanthan gum confirm the decrease in the permeability coefficient after biostabilisation, although this reduction is less significant with the use of the biopolymer obtained directly from the bacteria. Moreover, during 28 days of curing time the coefficient of permeability of the soil stabilized with commercial xanthan gum is approximately constant, while an increase in the permeability over time is seen with the xanthan-like strain, which may indicate some degradation of this type of gum.

Another recent study has shown that the mixture of 3% xanthan gum with a kaolin clay also decreases the gas permeability of the biostabilized soil with pure clay by up to two orders of magnitude, which is due to pore-clogging [35]. Moreover, the decrease in the permeability caused by xanthan gum is more significant with increasing water content, since more water is available to hydrate the biopolymers and to promote the pore-clogging [35].

Although the majority of the works published show that the biostabilization of soils with xanthan gum induces a reduction in the permeability coefficient, Biju and Arnepalli's results [46] are not in line with this tendency; in fact, the mixture of xanthan gum (0.5–3%) with sand–bentonite showed a slight increase in the permeability, which is justified by the aggregation of clay platelets, which induces wider effective flow paths [46].

Analysis of SEM images of a soil–xanthan gum mixture from different curing times [29] shows that shrinkage occurs during the time of the crosslinking of the xanthan gum within the matrices formed in the soil. This consequently increased the voids in the soil, as well as their permeability. In general, it is considered that the final stage of shrinking in the biopolymers is the transition phase between the rubber gel and the vitreous state. Through this transformation, the biopolymer gains more strength, which was also verified in the shear strength tests, and further, the volume of biopolymer decreased inside the empty spaces of the soil, which also generated an increase in its permeability.

4.3.3. Stability and Durability

Another important concern that must be addressed when using xanthan gum is related to durability. It is considered to be the case that all types of polymers, even those that are petroleum or biological-based, are biodegradable in nature. However, the rate of degradation changes according to its type of composition. So, polymers that degrade slowly or at slow rates may be considered non-biodegradable or durable [27]. Thus, the long term durability conditions for a biopolymer treated soil must be considered against its bio-decomposition, specifically under humidification and drying cycles. These conditions should be studied for each case and confirmed to provide safety recommendations concerning the use of biopolymers used in Geotechnical Engineering [30].

Xanthan gum, which is considered a very stable biopolymer, is resistant to different conditions such as thermal decomposition below 250 °C, oxidation, acid/alkaline environments and high concentrations of salt [18]. Results from Muguda et al. [41] showed that the stabilization of a soil composed of sand, gravel and kaolin with 1% xanthan gum has a satisfactory durability performance against water-induced deterioration and good hygroscopic performance. Moreover, the results of slake durability carried out with xanthan gum-treated sand indicated that the resistance to disintegration upon interaction with water of the treated soil is stronger than that of the sand stabilized with Portland cement [33].

The issue of durability in the case of biopolymers applied as soil improvement materials should be evaluated in more detail on a case-by-case basis and knowledge about this subject would thus improve with more research in this area. Although it can be considered that the use of xanthan gum has numerous benefits for soil improvement, it is important to mention that the interaction between this biopolymer and the soil depends on some conditions, such as soil type, soil–biopolymer interactions or temperature and reaction with water. In general, it is expected that the physical interaction and chemical properties of xanthan gum will be better understood, in the future, so that it can be appropriately applied to the soil type and purpose according to the relevant geotechnical characteristics [22].

4.3.4. Other Properties

Relative to the density of the sandy soil, the results reveal an increase in density with the increment of the xanthan gum concentration [29]. The increasing density can be explained based on the nature of xanthan gum, which reduces the friction between the particles, thus inducing a better arrangement of the soil particles. This viscosity can also be expected to increase as the molecular weight of the biopolymer increases because a crystallization in its macromolecule chain is more likely. This consequently leads to an increase in the degree of crosslinking within the soil matrix.

In some works, it was observed that the optimum water content (OWC) increased with higher gum concentrations for all samples and this is due to the increase in the amount of water absorbed with the increasing concentration of the biopolymer [21,24,26].

Some experiments have shown that the use of xanthan gum also changes the liquid limit (LL) of the biostabilised soil, and this effect depends on of two opposite effects, the formation of a viscous hydrogel in the voids and the aggregation of clay particles. When the latter effect is dominant, an increase in the LL is obtained for a higher content of xanthan gum. This was obtained with a clayed silt-sand soil, montmorillonite clay [25], a soft marine soil [24] and a mine tailing [38]. For a kaolinite clay, the LL has a peak for a low content of xanthan gum (close to 0.5%), followed by its decrease. Thus, during the first stage, the formation of the hydrogel is more significant, while after the peak, aggregation is the key factor [25,47].

5. Conclusions and Future Research

In this review, we have summarized the importance of the application of biopolymers in geotechnical engineering and have provided detailed examples of the applications of the xanthan gum biopolymer in different types of soil. Recent studies have shown improvement in soil strength induced by adding biopolymers and, in addition, some advantages have been noted concerning the use of xanthan gum in relation to other types of biopolymers. Through the bioclogging method, the use of xanthan gum allows the soil pores to be filled and forms a barrier capable of increasing the strength and cohesion, and reducing the permeability coefficient, of the soil. Although the use of xanthan gum has benefits for both soil improvement and stabilization, it is important to ensure that the right conditions exist for a properly controlled and efficient soil–biopolymer interaction.

In conclusion, the practical application of these studies is essential to prove the results discussed here and, for this reason, future studies should focus on tests that are able to prove that the xanthan gum–soil mixture improves the soil's properties, since xanthan gum's potential is still relatively understudied.

Author Contributions: Conceptualization, A.C.P., P.V.M. and A.P.C.; writing—original draft preparation, A.M., A.C.P. and P.V.M.; writing—review and editing, A.M., P.V.M., P.V.O. and A.P.C.; funding acquisition, P.V.M. and P.V.O. All authors have read and agreed to the published version of the manuscript.

Funding: This work was supported by the projects PTW PTDC/AAGREC/3839/2014 and POCI-01-0145-FEDER-031820 funded by the COMPETE2020, and by Fundação para a Ciência e Tecnologia (FCT), project PTDC/CTA-AMB/31820/2017 and FCT UIDB/00285/2020, BIORECOVER H2020 grant agreement 821096, project POCI-01-0145-FEDER-028382 and the R&D Unit Institute for Sustainability and Innovation in Structural Engineering (ISISE), under reference UIDB/04029/2020.

Institutional Review Board Statement: Not applicable.

Informed Consent Statement: Not applicable.

Data Availability Statement: No new data were created or analyzed in this study. Data sharing is not applicable to this article.

Acknowledgments: This work had the lab technician support of J.A. Lopes, Department of Civil Engineering, University of Coimbra.

Conflicts of Interest: The authors declare no conflict of interest.

References

1. Chang, I.; Im, J.; Cho, G.-C. Introduction of microbial biopolymers in soil treatment for future environmentally friendly and sustainable geotechnical engineering. *Sustainability* **2016**, *8*, 251. [CrossRef]
2. Ivanov, V.; Chu, J. Applications of microorganisms to geotechnical engineering for bioclogging and biocementation of soil in situ. *Rev. Environ. Sci. Biotechnol.* **2008**, *7*, 139–153. [CrossRef]
3. Murphy, E.M.; Ginn, T.R. Modeling microbial processes in porous media. *Hydrogeol. J.* **2000**, *8*, 142–158. [CrossRef]
4. Basu, D.; Misra, A. Sustainability in Geotechnical Engineering. In Proceedings of the 18th International Conference on Soil Mechanics and Geotechnical Engineering, Paris, France, 2–6 September 2013. [CrossRef]
5. Narjary, B.; Aggarwal, P.; Singh, A.; Chakraborty, D.; Singh, R. Water availability in different soils in relation to hydrogel application. *Geoderma* **2012**, *187–188*, 94–101. [CrossRef]
6. Murthy, V.N.S. *Geotechnical Engineering: Principles and Pratices of Soil Mechanics and Foudation Engineering*; Marcel Dekker, Inc.: New York, NY, USA, 2016.
7. Baveye, P.; Vandevivere, P.; Blythe, L.H.; DeLeo, P.; Sanchez, L.D. Environmental Impact and Mechanisms of the Biological Clogging of Saturated Soils and Aquifer Materials. *Crit. Rev. Environ. Sci. Technol.* **2010**, *28*, 123–191. [CrossRef]
8. Velde, K.V.D.; Kiekens, P. Biopolymers: Overview of several properties and consequences on their applications. *Polym. Test.* **2002**, *21*, 433–442. [CrossRef]
9. Choi, S.; Chang, I.; Lee, M.; Lee, J.; Han, J.; Kwon, T. Review on geotechnical engineering properties of sands treated by microbially induced calcium carbonate precipitation (MICP) and biopolymers. *Constr. Build. Mater.* **2020**, *246*, 118415. [CrossRef]
10. Jose, A.; Carvalho, F. Starch as Source of Polymeric Materials. In *Biopolymers: Biomedical and Environmental Applications*; Wiley: Salem, MA, USA, 2011.
11. Khachatoorian, R.; Petrisor, I.G.; Khan, C.C.; Yen, T.F. Biopolymer plugging effect: Laboratory-pressurized pumping flow studies. *J. Petroleum. Sci. Eng.* **2013**, *38*, 13–21. [CrossRef]

12. Smitha, S.; Sachan, A. Use of agar biopolymer to improve the shear strength behavior of sabarmati sand. *Int. J. Geotech. Eng.* **2016**, *10*, 387–400. [CrossRef]
13. Mataix-Solera, J.; Guerrero, C.; Hernandez, M.T.; Garcıa-Orenes, F.; Mataix-Beneyto, J.; Gomez, I.; Escalante, B. Improving soil physical properties related to hydrological behaviour by inducing microbiological changes after organic amendments. *Geophys. Res.* **2005**, *7*, 00341. [CrossRef]
14. Chang, I.; Prasidhi, A.K.; Im, J.; Shin, H.D.; Cho, G.C. Soil treatment using microbial biopolymers for anti-desertification purposes. *Geoderma* **2016**, *44*, 253–254. [CrossRef]
15. Chang, I.; Prasidhi, A.K.; Im, J.; Cho, G.C. Soil strengthening using thermo-gelation biopolymers. *Const. Build. Mat.* **2015**, *77*, 430–438. [CrossRef]
16. Mitchell, J.K.; Santamarina, J.C. Biological considerations in geotechnical engineering. *J. Geotech. Geoenviron. Eng.* **2005**, *131*, 1222–1233. [CrossRef]
17. Chang, I.; Im, J.; Prasidhi, A.K.; Cho, G.C. Effects of Xanthan gum biopolymer on soil strengthening. *Constr. Build. Mater.* **2015**, *74*, 65–72. [CrossRef]
18. Yakimets, I.; Paes, S.S.; Wellner, N.; Smith, A.C.; Wilson, R.H.; Mitchell, J.R. Effect of Water Content on the Structural Reorganization and Elastic Properties of Biopolymer Films: A Comparative Study. *Biomacromolecules* **2007**, *8*, 1710–1722. [CrossRef]
19. Stewart, T.L.; Fogler, H.S. Biomass plug development and propagation in porous media. *Biotechnol. Bioeng.* **2001**, *72*, 353–363. [CrossRef]
20. Garcia-Ochoa, F.; Santos, V.E.; Casas, J.A.; Gomez, E. Xanthan gum: Production, recovery, and properties. *Biotechnol. Adv.* **2000**, *18*, 549–579. [CrossRef]
21. Bouazza, A.; Gates, P.W.; Ranjith, P.G. Hydraulic conductivity of biopolymer-treated silty sand. *Geotechnique* **2009**, *59*, 71–72. [CrossRef]
22. Mendonça, A.C.S. Use of Microbial Biopolymer to Decrease Soil Permeability by Bioclogging. Master's Thesis, University of Coimbra, Coimbra, Portugal, 2020.
23. Pansu, M.; Gautheyrou, J. *Handbook of Soil Analysis—Mineralogical, Organic and Inorganic Methods*; Springer-Verlag: Berlin, Germany, 2006.
24. Kwon, Y.M.; Chang, I.; Lee, M.; Cho, G.C. Geotechnical engineering behavior of biopolymer-treated soft marine soil. *Geomech. Eng.* **2019**, *17*, 453–464. [CrossRef]
25. Chang, I.; Kwon, Y.-M.; Im, J.; Cho, G.C. Soil consistency and interparticle characteristics of Xanthan gum biopolymer–containing soils with pore-fluid variation. *Can. Geotech. J.* **2019**, *56*, 1206–1213. [CrossRef]
26. Kucharski, E.S.; Winchester, W.; Leeming, W.A.; Cord-Ruwisch, R.; Muir, C.; Banjup, W.A.; Whiffin, V.S.; Al-Thawadi, S.; Mutlaq, J. Microbial Biocementation. International Patent Application No.PCT/AU2005/001927, 29 June 2006.
27. Krishna Leela, J.; Sharma, G. Studies on xanthan production from *Xanthomonas campestris*. *Bioprocess. Eng.* **2000**, *23*, 687–689. [CrossRef]
28. Jang, J. A Review of the Application of Biopolymers on Geotechnical Engineering and the Strengthening Mechanisms between Typical Biopolymers and Soils. *Adv. Mater. Sci. Eng. Hindawi* **2020**, *2020*, 1–16. [CrossRef]
29. Ayeldeen, K.M.; Abdelazim, M.N.; Mostafa, A.E.S. Evaluating the physical characteristics of biopolymer/soil mixtures. *Arab. J. Geosci.* **2016**, *9*, 371. [CrossRef]
30. Lee, S.; Chang, I.; Chung, M.K.; Kim, Y.; Kee, J. Geotechnical shear behavior of xanthan gum biopolymer treated sand from direct shear testing. *Geomech. Eng.* **2017**, *12*, 831–847. [CrossRef]
31. Cabalar, A.F.; Wiszniewski, M.; Skutnik, Z. Effects of Xanthan Gum Biopolymer on the Permeability, Odometer, Unconfined Compressive and Triaxial Shear Behavior of a Sand. *Soil Mech. Found. Eng.* **2017**, *54*, 356–361. [CrossRef]
32. Cabral, D.J.R. Stabilization of a Soil through the Use of Biopolymers: Effect of the Content and Curing Time. Master's Thesis, University of Coimbra, Coimbra, Portugal, 2020. (In Portuguese)
33. Qureshi, M.U.; Chang, I.; Al-Sadarani, K. Strength and durability characteristics of biopolymer-treated desert sand. *Geomech. Eng.* **2017**, *12*, 785–801. [CrossRef]
34. Lee, S.; Chung, M.; Park, H.M.; Song, K.-I.; Chang, I. Xanthan gum Biopolymer as Soil-Stabilization Binder for Road Construction Using Local Soil in Sri Lanka. *J. Mater. Civ. Eng.* **2019**, *31*, 06019012. [CrossRef]
35. Joga, J.R.; Varaprasad, B.J.S. Sustainable Improvement of Expansive Clays Using Xanthan Gum as a Biopolymer. *Civ. Eng. J.* **2019**, *5*, 1863–1903. [CrossRef]
36. Chang, I.; Jeon, M.; Cho, G.C. Application of microbial biopolymers as an alternative construction binder for earth buildings in underdeveloped countries. *Int. J. Polym. Sci.* **2015**. [CrossRef]
37. Fatehi, H.; Abtahi, S.M.; Hashemolhosseini, H.; Hejazi, S.M. A novel study on using protein based biopolymers in soil strengthening. *Constr. Build. Mater.* **2018**, *167*, 813–821. [CrossRef]
38. Chen, R.; Zhang, L.; Budhu, M. Biopolymer Stabilization of Mine Tailings. *J. Geotech. Geoenviron. Eng.* **2013**, *139*, 1802–1807. [CrossRef]
39. Latifi, N.; Horpibulsuk, S.; Meehan, C.L.; Abd Majid, M.Z.; Tahir, M.M.; Mohamad, E.T. Improvement of Problematic Soils with Biopolymer—An Environmentally Friendly Soil Stabilizer. *J. Mater. Civ. Eng.* **2017**, *29*, 04016204. [CrossRef]
40. Soldo, A.; Miletić, M.; Auad, M.L. Biopolymers as a sustainable solution for the enhancement of soil mechanical properties. *Sci. Rep.* **2020**, *10*, 1–13. [CrossRef] [PubMed]

41. Muguda, S.; Booth, S.J.; Hughes, P.N.; Augarde, C.E.; Perlot, C.; Bruno, A.W.; Gallipoli, D. Mechanical properties of biopolymer-stabilised soil-based construction materials. *Géotech. Lett.* **2017**, *7*, 309–314. [CrossRef]
42. Chen, C.; Wu, L.; Perdjon, M.; Huang, X.; Peng, Y. The drying effect on xanthan gum biopolymer treated sandy soil shear strength. *Constr. Build. Mater.* **2019**, *197*, 271–279. [CrossRef]
43. Czarnes, S.; Hallett, P.D. Root and microbial-derived mucilages affect soil structure and water transport. *Eur. J. Soil Sci.* **2000**, *51*, 435–443. [CrossRef]
44. Khatami, H.; O'Kelly, B. Improving mechanical properties of sand using biopolymers. *J. Geotech. Geoenviron.* **2013**, *139*, 1402–1406. [CrossRef]
45. Dehghan, H.; Tabarsa, A.; Latifi, N.; Bagheri, Y. Use of Xanthan and guar gums in soil strengthening. *Clean Technol. Environ. Policy* **2019**, *21*, 155–165. [CrossRef]
46. Biju, M.S.; Arnepalli, D.N. Effect of biopolymers on permeability of sand-bentonite mixtures. *J. Rock Mech. Geotech. Eng.* **2020**, *12*, 1093–1102. [CrossRef]
47. Nugent, R.; Zhang, G.; Gambrell, R. Effect of exopolymers on the liquid limit of clays and its engineering implications. *Transp. Res. Rec.* **2010**, *2101*, 34–43. [CrossRef]

Article

Effect of Aeration Mode on Microbial Structure and Efficiency of Treatment of TSS-Rich Wastewater from Meat Processing

Piotr Jachimowicz, Agnieszka Cydzik-Kwiatkowska * and Patrycja Szklarz

Department of Environmental Biotechnology, University of Warmia and Mazury in Olsztyn, 10-709 Olsztyn, Poland; piotr.jachimowicz@uwm.edu.pl (P.J.); patrycja.szklarz@student.uwm.edu.pl (P.S.)

* Correspondence: agnieszka.cydzik@uwm.edu.pl

Received: 26 August 2020; Accepted: 18 October 2020; Published: 22 October 2020

Abstract: The present study investigated the effect of aeration mode on microbial structure and efficiency of treatment of wastewater with a high concentration of suspended solids (TSS) from meat processing in sequencing batch reactors (R). R_1 was constantly aerated, while in R_2 intermittent aeration was applied. DNA was isolated from biomass and analyzed using next-generation sequencing (NGS) and real-time PCR. As a result, in R_1 aerobic granular sludge was cultivated (SVI_{30} = 44 mL g^{-1} MLSS), while in R_2 a very well-settling mixture of aerobic granules and activated sludge was obtained (SVI_{30} = 65 mL g^{-1} MLSS). Intermittent aeration significantly increased denitrification and phosphorus removal efficiencies (68% vs. 43%, 73% vs. 65%, respectively) but resulted in decomposition of extracellular polymeric substances and worse-settling properties of biomass. In both reactors, microbial structure significantly changed in time; an increase in relative abundances of *Arenimonas* sp., *Rhodobacterace*, *Thauera* sp., and *Dokdonella* sp. characterized the biomass of stable treatment of meat-processing wastewater. Constant aeration in R_1 cycle favored growth of glycogen-accumulating *Amaricoccus tamworthensis* (10.9%) and resulted in 2.4 times and 1.4 times greater number of ammonia-oxidizing bacteria and full-denitrifiers genes in biomass, respectively, compared to the R_2.

Keywords: intermittent aeration; TSS-rich meat-processing wastewater; extracellular polymeric substances; complete denitrification; microbial structure

---

1. Introduction

The meat-processing industry uses about 24% of freshwater consumed by the food and beverage industry and up to 29% of that consumed by the agricultural sector worldwide [1]. About 341.2 million tons of meat were produced globally in year 2018 (63.9 million tons in Europe) [2]. These numbers will increase because world beef, pork, and poultry production are projected to double by 2050 [3]. Meat-processing effluents contain a lot of fats, fibers, proteins, and pathogens but also pharmaceuticals and detergents used for veterinary and cleaning activities [4]. Fats and proteins are main components of the particulate matter in meat-processing wastewater constituting on average 55.3% and 27.1% of dry mass, respectively [5]. Fat and proteins in meat industry wastewaters are usually poorly biodegradable that results in generation of odors, foam, and poor flocculation and settling of biomass that decreases the efficiency of biological processes. Oil and grease adsorb on cell surfaces lowering transfer aqueous phase and pollutant conversion rates [6].

One method that shows promise for treating this difficult-to-treat effluent is the use of granular sludge technology. Granular sludge is a self-immobilized consortium, densely packed with microorganisms used in wastewater treatment reactors [7]. The volumes of reactors with aerobic granular sludge can be smaller than those with activated sludge, because of very good settling properties of granules and a high concentration of microorganisms in their structure. During formation of granules microbial competition and changes in cellular metabolism occur that determine granule morphology and treatment efficiency [8,9].

Granulation depends on numerous factors such as sequencing batch regimes, feeding characteristics, or organic and nitrogen loadings. The presence of aerobic, anoxic and anaerobic zones in the granule structure provides favorable conditions for the growth of both aerobic nitrifiers, and anaerobic and anoxic microorganisms. Granulation process is faster in plug-flow anaerobic feeding reactor than in fully aerobic reactor; however, the granular biomass formed in the fully aerobic configuration is more stable mostly due to better retention of nitrifying organisms favoring nitrogen removal efficiency [10]. Positive effect of anoxic or anaerobic phases in the cycle of granular sequencing batch reactor (GSBR) on nitrogen removal was also observed in other studies [11,12]. There are reports informing that aerobic granular sludge can be successfully applied to treat slaughterhouse wastewater in a lab-scale GSBR [13,14], but studies connecting the efficiency of treatment with in depth analysis of structure of microbial community using next-generation sequencing (NGS) in granules are lacking.

The presence of extracellular polymeric substances (EPS) has a significant influence on the physicochemical properties of microbial biomass aggregates, including surface change, structure, flocculation, and settling properties [15]. The aeration conditions influence the production of EPS and it was observed that the content of EPS in biomass decreases under anaerobic conditions [16]. Content of carbohydrates in EPS from biomass from reactors with a high dissolved oxygen level increased with time, whereas the protein content remained unchanged [17]. The protein present in high concentrations in wastewater from meat industry can be incorporated in the bound EPS in biomass [18].

Therefore, the aim of this study was to explore interrelationships between the aeration mode, microbial community composition, and efficiency of nitrogen, phosphorus, organic compounds (COD), and total suspended solids removal in batch reactors treating wastewater from meat-processing industry with a very high concentration of suspended solids. The microbial groups involved in pollutant degradation in constantly and intermittently aerated granules treating TSS-rich meat-processing wastewater were indicated.

2. Materials and Methods

2.1. Experimental Design

Industrial wastewater (Table 1) was obtained from a meat-processing plant in Morliny (Poland) after pre-treatment in a flotation tank. During transport wastewater was stored in ice. The experiment was conducted in two batch reactors with aerobic granular sludge (GSBR) at a room temperature (25 °C). The reactor (Figure 1) had a height to diameter (H/D) ratio of 1.6 and a working volume of 10 L. Volumetric exchange rate of 20% cycle^{-1} and a cycle length of 6 h was maintained by programmable logic controllers. Air was supplied via aeration grids in the bottom of the reactors. During aeration phases, the up-flow superficial air velocity was 0.5 cm s^{-1} and the free dissolved oxygen concentration reached saturation. The reactors were inoculated with aerobic granular sludge (AGS) cultivated in a full-scale wastewater treatment plant in Lubawa [19]. The reactor cycle (Figure 1) consisted of 10 min of settling, 30 min of filling and decantation, and 320 min of reaction phase. R_1 was constantly aerated, while R_2 was intermittently aerated, with 30 min non-aerated phases, after every 60 min of aeration. The operational parameters of the reactors favored formation of aerobic granules.

During reactor operation, pollutants concentration was analyzed both in the influent and the effluent. Additionally, to determine the variance of pollutant concentrations within the reactor cycle. Measurements were performed in duplicate every half hour during the period of stable reactor performance. The experiment was conducted for 200 cycles. For efficiency calculations, the technological results from the last 50 cycles of stable reactor operation were used.

Table 1. The concentrations of pollutants in wastewater from meat processing (COD—Chemical oxygen demand; BOD—Biochemical oxygen demand; TKN—Total Kjeldahl nitrogen; TP—Total phosphorus; VFA—Volatile fatty acids).

Parameter	Unit	Value
COD	mg L^{-1}	1686.8 ± 612.0
BOD	mg L^{-1}	1268.4 ± 421.5
TKN	mg L^{-1}	69.8 ± 19.5
N-NH_4^+	mg L^{-1}	20.4 ± 13.8
TP	mg L^{-1}	23.9 ± 4.3
P-PO_4^{3-}	mg L^{-1}	10.3 ± 14.1
Fats	mg L^{-1}	450 ± 37
VFA	mg L^{-1}	910 ± 154.2
Acetic acid	mg L^{-1}	254 ± 24.6
Alkalinity	-	9 ± 0.3
pH	-	8.0 ±0.2

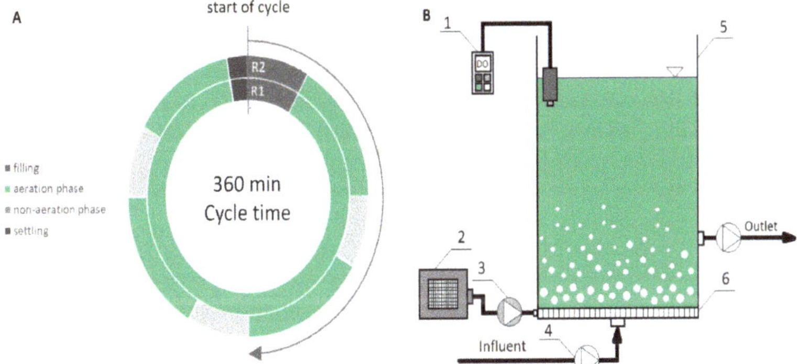

Figure 1. Organization of reactors cycle (**A**) and a scheme of the reactors: 1—Oxygen detector, 2—Compressor, 3—Air pump, 4—Peristaltic pump, 5—Reactor, 6—Fine bubble diffuser (**B**).

2.2. Analytical Methods

The reactors wastewater and biomass were analyzed in accordance with APHA [20]; the sludge volumetric index was measured after 30 min of settling (SVI_{30}). The oxygen concentration in the reactor was measured using a YSI ProODO™ probe (YSI). Chemical analyzes of the influent and effluent from wastewater included COD, BOD, VFA, N-NH_4, N-NO_2, N-NO_3, and P-PO_4. Alkalinity, and pH were measured using TitroLine easy (Donserv). All analyses were conducted in triplicate.

Granule morphology and size was evaluated using wet sieving (water temperature 12 °C) in an AS 200 screening unit (Retsch) using sieves with mesh sizes of 1 mm, 710 μm, 500 μm, 355 μm, 125 μm, 90 μm and 40 μm. Sieving lasted for 10 min with a vibration amplitude of 50 mm.

Soluble (Sol-EPS), loosely bound (LB-EPS) and tightly bound (TB-EPS) extracellular polymeric substances (EPS) were isolated from the biomass collected from the reactors at the beginning (25 min) and end (269 min) of the cycle as described in Rusanowska et al. [21]. Concentration of proteins (PN) and polysaccharides (PS) in particular fractions of EPS was measured according to Frølund et al. [22], and Lowry methods, respectively.

2.3. Next-Generation Sequencing

For molecular analyses of the granular sludge microbiome, sludge from the two reactors was sampled at the beginning of the experiment and after 56, 128, and 164 cycles and stored at −20 °C. DNA was isolated in triplicate using a FastDNA®SPIN kit for soil (MPBiomedicals, Irvine, CA, USA) and its purity and concentration was measured using a NanoDrop spectrometer (Thermo Scientific, Waltham, MA, USA). Then the DNA isolated from these three samples was pooled together for further analyses. The 939F/1492R primer set (5′-TTGACGGGGGCCCGCAC-3′/5′-TACCTTGTTACGACTT-3′) was used to amplify the V6 and V8 regions of the bacterial *16S rRNA* gene. Sequencing of amplicons using the Illumina MiSeq platform was performed at Research and Testing Laboratory (USA). Sequencing results were deposited in the Sequence Read Archive (SRA, BioProject PRJNA613085). Depending on the sample, from 7237 to 26594 sequences were obtained (Appendix A Table A1). The sequences were analyzed bioinformatically as described in Świątczak et al. [23]. To characterize microbial diversity at the species level, the Shannon-Wiener (H′) index of diversity was calculated [24].

2.4. Real-Time PCR

Real-time PCR was performed to assess the absolute abundance of total bacteria (*16S rRNA* gene), ammonia-oxidizing bacteria (*amoA* gene), and denitrifiers (*nosZ* gene) in aerobic granules. Standard curves for analysis of the copy number of the *amoA* and *nosZ* genes were constructed using a plasmid with a cloned insert (PCR product of known size) using the Clone Jet PCR cloning kit (Thermo Scientific) and freshly prepared, chemically competent *Escherichia coli* 109 (Promega) for cloning. In all clones the presence of the insert was confirmed by PCR with appropriate primers (Appendix A Table A1). To construct a standard curve for absolute quantification of total bacteria, DNA isolated from *E. coli* 109 (Promega) was used. The reaction mixture for real-time PCR contained 5 ng/μL of the template DNA, 10 μL of Maxima SYBR Green/ROXqPCR Master Mix (2×) (Thermo Scientific, Waltham, MA, USA), primers (concentrations are given in Appendix A Table A2) and ultrapure water for a final volume of 20 μL. The amplification reaction began with 2 min at 50 °C and 10 min at 95 °C; then 40 cycles of amplification proceeded according to the thermal profiles given in Appendix A Table A2. Reactions were carried out in a 7500 Real-Time PCR System (Applied Biosystems) and data were analyzed with Sequence Detection Software, version 1.3 (Applied Biosystems). Reactions were normalized by adding the same amount of DNA to each reaction tube. To confirm reaction specificity, the positive and negative controls were performed. To confirm the melting temperature of PCR products, dissociation step and agarose electrophoresis in the presence of a molecular marker were performed.

2.5. Statistics

For statistical analysis, STATISTICA 13.1 software (StatSoft) was used. A value of $p \leq 0.05$ was defined as significant.

Normality and the homogeneity of variance were examined using the Shapiro–Wilk and the Levene's tests, respectively. If these assumptions were correct, Student's t-test was used to test the differences between the means. Technological results and abundance of microorganisms were correlated using Pearson's correlation coefficient. Links between technological parameters (cycle number and the number of anoxic phases in the reactor cycle) and microbial structure were assessed with Canonical Correspondence Analysis (CCA) with Monte Carlo permutation testing (F statistics, 499 permutations). The analyses were conducted using CANOCO for Windows ver. 5.0 and CANODRAW [25].

3. Results and Discussion

The obtained results indicate that in the constantly aerated reactor, aerobic granule formation was favored, whereas in the intermittently aerated reactor, a mixture of aerobic granules and activated sludge flocks was obtained. The average biomass concentration in R_1 was 4.6 ± 0.9 g MLSS L^{-1} and that in R_2 was 3.4 ± 1.5 g MLSS L^{-1}, but the concentration of TSS in the effluent was nearly 2 times higher in R_2 than in R_1 (0.20 g L^{-1} TSS vs. 0.11 g L^{-1} TSS). The higher concentration of biomass in R_2 may have resulted from organics being used, not for biomass growth, but for more efficient denitrification and phosphorus removal than in R_1. The SVI_{30} was higher in R_2 (65 mL g^{-1} MLSS) than in R_1 (44 mL g^{-1} MLSS). Aerobic granules exhibit low SVI values (<50 mL/g MLSS), which is in sharp contrast to activated sludge, in which SVI values exceed 120 mL/g MLSS [9].

In the biomass from R_1, granules with sizes from 90 to 125 μm predominated, constituting about 25.1% of the biomass, and granules with diameters over 1 mm were observed (Figure 2). In R_2, about 50.0% of the biomass consisted of granules smaller than 90 μm. Both the superficial air velocity and the oxygen concentration affect aerobic granulation [26]. Tay et al. [27], reported that a higher shear force resulting from a higher superficial air velocity supported formation of more compact, denser, and stronger aerobic granules. The relatively low air velocity in the present experiment (0.5 cm s^{-1}) and the reduced oxygen concentration in R_2 worsened the settling properties of the biomass and precluded full granulation of the biomass. The size and formation of stable aerobic granules correlates with the efficiency of pollutant removal because low air velocity favors selection of slow-growing organisms [28]. In a study involving the treatment of slaughterhouse wastewater, the size of granules ranged from 0.6 to 1.2 mm, which favored a high abundance of ammonium-oxidizing bacteria [13]. In the present study, the larger granule diameters in R_1 explain the greater abundance of nitrifiers in this biomass, as indicated by *amoA* gene analysis, and the higher nitrification rate. A study by Luo et al. [29], revealed that larger granules possess a higher microbial biodiversity than smaller granules. On the one hand, increasing granule size decreases the surface/volume ratio, resulting in a higher surface loading of pollutants, which may affect aerobic transformations in the biomass. In larger granules, the depth to which oxygen permeates will be less than in smaller granules, and larger anaerobic/anoxic zones will be present, which provide spaces for the growth of denitrifying bacteria. In contrast, in smaller granules, which have a larger surface/volume ratio, oxygen can permeate deeper than in larger granules and provide more aerobic zones for the growth of ammonia-oxidizing bacteria (AOB).

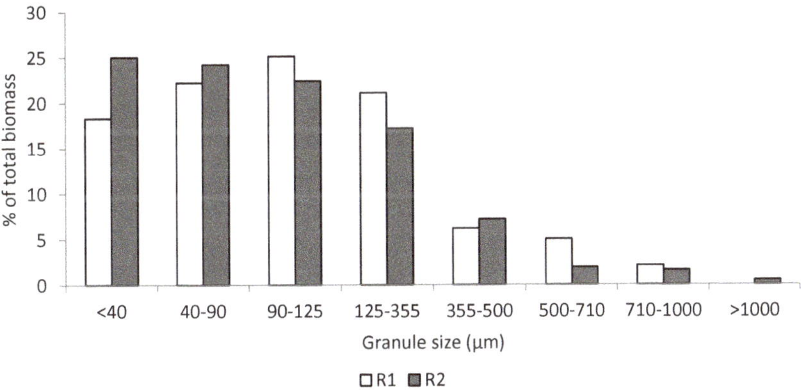

Figure 2. The distribution of granule particle sizes in the batch reactors.

In the present study, large changes in the structure of microorganisms colonizing the inoculum and biomass (Figure 3) were noted throughout the experiment. Weissbrodt et al. [30] also observed that during the granulation of activated sludge, the percentage of *Competibacter* sp. increased, and bacteria of the *Xanthomonadaceae*, *Rhodospirillaceae*, and *Aminobacter* families appeared. According to those authors, the occurrence of bacteria such as *Zoogloea* sp., *Xanthomonadaceae*, *Sphingomonadales, and Rhizobiales* promotes the formation of EPS, which are responsible for biomass granulation. In the present study, *Dokdonella* sp. were found to be responsible for the formation of EPS and denitrification in biomass. The presence of microorganisms from the above-mentioned families promotes granulation and affects the strength of the resulting granules.

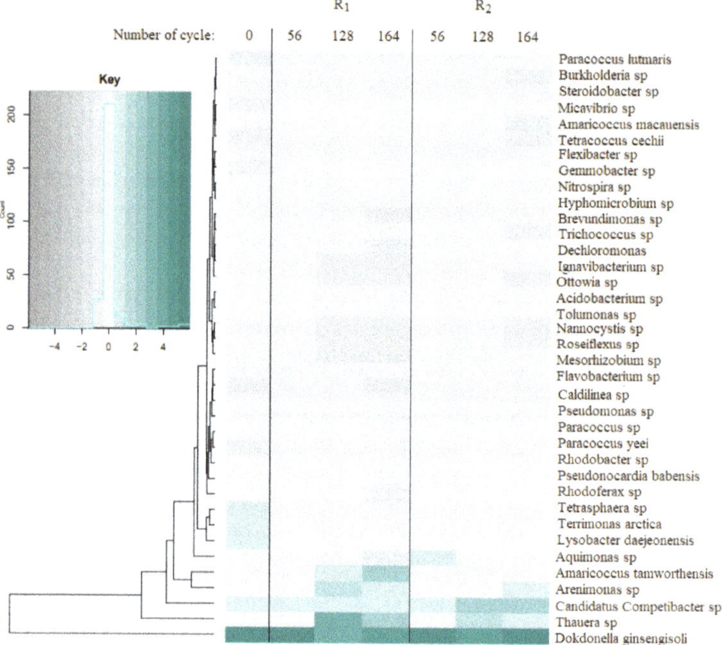

Figure 3. Heatmap of relative abundance of bacterial taxa present in aerobic granules (abundance higher that 0.5%).

The quality of the treated wastewater in the present study indicates that systems with aerobic granules can successfully treat TSS-rich wastewater from the meat industry. Organic compounds were removed with a very high efficiency in both reactors (above 94%). This indicates that periodic decreases in oxygen concentration during the reactor cycle, which resulted from intermittent aeration, did not reduce the efficiency of organics removal.

The average COD concentrations were 91.6 ± 13.5 mg L^{-1} and 115.6 ± 43.6 mg L^{-1} in the outflows from R_1 and R_2, respectively, but COD removal was more stable in R_1 (Figure 4). The COD removal rate was similar in both reactors (197.1 mg (L h^{-1}) vs. 194.6 mg (L h^{-1}) in R_1 and R_2, respectively). The carbon source that is available in wastewater affects the composition of the microbiological community. De Sousa Rollemberg et al. [31], observed that in a reactor to which acetate was dosed, microorganisms from the genera *Flavobacterium* (30%), *Thauera* (15%) and *Aquimonas* (12%) dominated. In a reactor to which ethanol was dosed, *Paracoccus* sp. (36%), *Fluviicola* sp. (20%) and *Sediminibacterium* sp. (9%) dominated, while addition of glucose resulted in the growth of numerous polyphosphate-accumulating organisms (PAOs) belonging to the taxa *Cryomorphaceae*, *Rhodobacteraceae*, *Cytophagaceae*, and *Saprospiraceae*. Acetate is the main source of carbon in wastewater from the meat industry, which may explain the high percentage of *Thauera* sp. (up to 20.1%) in the granules in the present study.

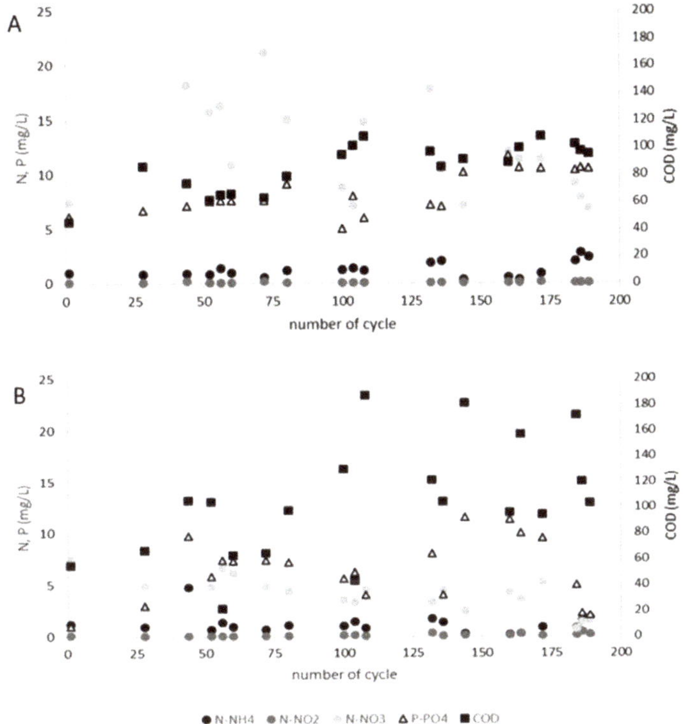

Figure 4. Changes in COD, nitrogen, and phosphorus concentrations in the effluents from the constantly aerated reactor, R_1 (**A**), and the reactor with intermittent aeration phases, R_2 (**B**).

The average N-NH_4 concentration in each effluent was around 1 mg L^{-1} (Figure 4). The efficiencies of ammonium nitrogen removal were 82 and 86% in R_1 and R_2, respectively, which is similar to the efficiency obtained by Ersan and Erguder [32], in a granular sludge batch reactor fed with wastewater with a COD/N ratio of 15:2 and operated with anoxic phases. The concentration of N-NO_2 in the effluents was low (0.03 ± 0.05 mg L^{-1} vs. 0.12 ± 0.12 mg L^{-1} in R_1 and R_2, respectively). The average concentration of N-NO_3 in the outflow from R_2 was significantly lower than that in the outflow from R_1 (3.6 ± 1.5 mg L^{-1} vs. 7.0 ± 3.2 mg L^{-1}). Correspondingly, denitrification efficiency was higher in R_2 than in R_1 (68% vs. 43%). The environmental conditions favored the growth of aerobic denitrifiers; for example, *Pseudomonas* sp. and *Paracoccus* sp., which can increase the efficiency of denitrification under aerobic conditions, were identified in the analyzed biomass [33].

In the present study, although the overall efficiency of ammonium nitrogen removal was very similar in both reactors, the ammonium removal rate was more than three times higher in the intermittently aerated reactor (9.4 mg/(L h^{-1})) than in the constantly aerated one (3.2 mg/(L h^{-1}). This indicates that ammonium nitrogen oxidation is stimulated by anoxic conditions. It has been reported that increasing the number of anaerobic phases in the reactor cycle increased the total number of bacteria (as shown by abundance of *16S rRNA* gene), the number of AOB bacteria involved in nitrification (*amoA* gene abundance) and the number of complete denitrifiers (*nosZ* gene abundance) [34]. In the present study, such increases were not observed; however, the share of such taxa as *Thauera* sp., *Dokdonella ginsengisoli*, and *Lysobacter daejeonensis*, which participate in nitrogen transformations, was higher in the reactor with intermittent aeration than in the one with constant aeration. In both reactors, low concentrations of N-NO_2 (below 1 mg L^{-1}) were only observed in the first hour of the cycle. In R_1, the N-NO_3 concentration increased during the first two hours of the working cycle, after which it remained at a few mg L^{-1}. In R_2, its concentration initially increased to 5.86 mg/(L h^{-1}), and then it was removed at different rates in subsequent non-aerated phases (22.5 mg (L h^{-1}), 15.7 mg (L h^{-1}), and 23.6 mg (L h^{-1}) in phases 1, 2, and 3, respectively) (Figure 5).

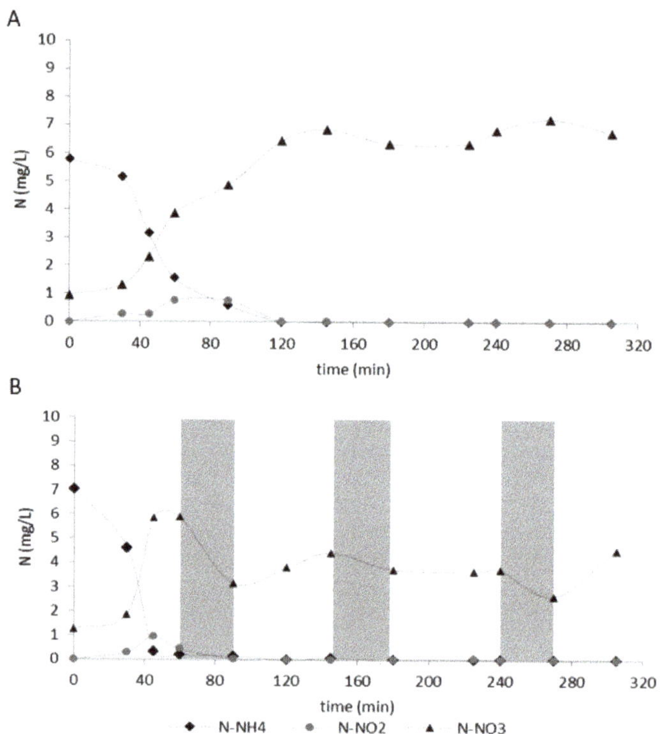

Figure 5. Concentration of ammonium, nitrite and nitrate during the cycles of (**A**) R_1 and (**B**) R_2. Gray shading indicates phases without aeration.

Due to intermittent aeration of R_2, the concentration of orthophosphates in the effluent from that reactor was significantly lower than that in the effluent from R_1 (6.5 ± 2.9 mg L^{-1} vs. 8.6 ± 1.7 mg L^{-1}, Figure 5). As a result, the removal efficiencies were 73% and 65%, respectively. A low oxygen concentration stimulates removal of COD, N and P due to heterotrophic growth inside granules consisting mostly of DPAO (denitrifying PAO) [35]. It has also been reported that lower phosphorus removal performance was often observed at high dissolved oxygen (DO) concentrations of 5.0 mg/L, while DO concentrations of approximately 2.5–3.0 mg/L seemed to be associated with the dominance of PAOs [36]. Taking into account the specific characteristics of wastewater from the meat industry and the lack of strict anaerobic phases, P removal was satisfactory and similar to the efficiencies obtained in other studies conducted in either constantly aerated granular batch reactors (83%) [14].

Presence of high concentrations of volatile fatty acids (VFA) stimulates PAOs to take up VFAs anaerobically and convert them to intracellular poly-β-hydroxyalkanoates (PHAs). PAOs gain the energy and reducing power required for anaerobic VFA uptake and conversion to PHA through the hydrolysis of their intracellularly stored polyphosphate (poly-P) and glycogen [37]. In the present study, PAO belonging to *Tetrasphaera* sp., *Arenimonas* sp., and *Pseudomonas* sp. [38,39], were identified and their presence was much higher in R_2. In R_1 percentage of *Amaricoccus tamworthensis* gradually increased. Falvo et al. [40], point out that *Amaricoccus* sp. in pure culture failed to synthesize poly-P aerobically but they could synthesize glycogen aerobically, they did not assimilate either acetate or glucose anaerobically, and PHAs synthesis occurred aerobically but not anaerobically. Glycogen-accumulating organisms (GAOs) are a bacterial group capable of competing with PAOs for VFA in wastewater. In the present study the most numerous GAO was *Candidatus Competibacter* that was present in both reactors [41].

The pH of the effluents was 8.4 and 8.2 in R_1 and R_2, respectively. The wastewater pH is considered to be a critical factor impacting N_2O emission. Hynes and Knowles [42] had pointed out that the maximum emission of N_2O was observed at pH 8.5 and decreased with decreasing pH, which can impact of *nosZ* gene abundance in the present study. Alkalinity of the effluent from R_1 (7.2 ± 0.5 mval L^{-1}) was significantly lower than from R_2 (to 8.5 ± 0.5 mval L^{-1}) pointing to more efficient recovery of alkalinity during denitrification in R_2 than in R_1. Even though the alkalinity was low, ammonium nitrogen was efficiently oxidized during the reactor cycle.

The distribution of EPS in the structure of granules plays an important role in improving the stability of the aerobic granular sludge and affects its SVI [43]. Extracellular polymers (EPS) were isolated from the aerobic granules at the beginning and end of the reactor cycle. The PN and PS concentrations in the three isolated EPS fractions are presented in Figure 6. In granules from R_1, content of PN and PS in Sol- and LB-EPS remained at a similar level in the GBSR cycle, but increased about 2 times at the end of the cycle in case of TB-EPS (Figure 6). In R_2 a significant decrease in PN concentration in Sol- and LB-EPS was observed at the end of the cycle that was accompanied by a slight increase in PN and PS concentrations in TB-EPS (Figure 6). PN and PS are the main components of EPS in aerobic granules. For granulation, the ratio of PN to PS in EPS is more important than their quantity in a given microbiological structure [44]. A high PN/PS ratio in EPS was observed in aerobic granules with very good settling properties [45]. Decreasing the amount of PN relative to PS decreases the consistency of granules, causes problems with biomass sedimentation and increase the risk of filamentous bacteria growth and sludge bulking [46]. It has been shown that the PN/PS ratio depends on the C/N ratio in the influent. For example, Zhang et al. [47], conducted research with C/N ratios of 5 and 15 in the wastewater fed to reactors. At the 15 C/N ratio, which was lower than C/N ratio used in our study (about 24), the PN/PS ratio in EPS was significantly lower than at the 5 C/N ratio, and granules with poorer settling and pollutant removal capacity were cultivated.

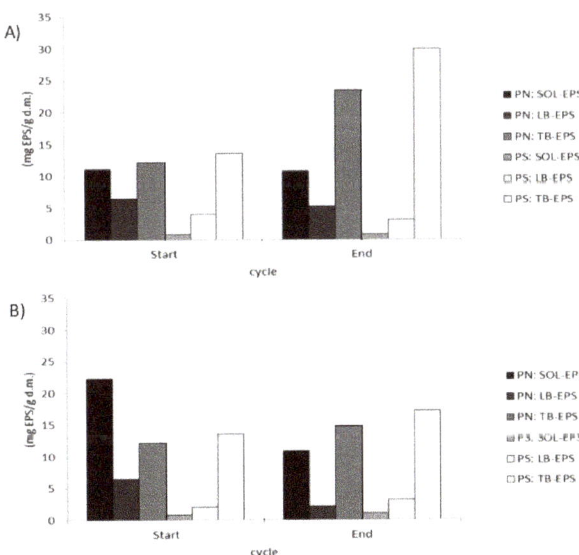

Figure 6. Content and composition of EPS in granules at the beginning and end of the reactor cycle in (**A**) R_1 and (**B**) R_2.

In the present study, the PN/PS ratio was low and remained at around 1 in both reactors at the beginning of the cycle. The largest changes in EPS concentration took place in the TB-EPS fraction. The PN and PS concentrations in TB-EPS at the beginning of the cycle was similar in both reactors but increased more than 1.5-time in R_1 in comparison with R_2 at the end of the cycle, which explains much better settling properties of granules from R_1. Higher concentration of TB-EPS at the end of R_1 cycle can be caused by the fact that in continuously aerated reactors, microorganisms use available carbon sources mainly for biomass and EPS synthesis. In intermittently aerated reactors, in contrast, they use more of the available carbon for denitrification and phosphorus removal. Up to 50% of the EPS in aerobic granular sludge can be used in the starvation phase and anaerobic conditions [48]. The high increase in PN in TB-EPS fraction at the end of the reactor cycle was advantageous for granules stability. Both TB-EPS and LB-EPS can support the aggregation of neighboring cells in AGS formation; however, LB-EPS does not bind cells together as strongly as TB-EPS and it can be more easily destroyed, which leads to the destruction of the granule structure [49].

Real-time PCR was used to estimate the total number of bacteria (*16S rRNA* gene copy number), the number of gene for *amoA* (ammonium monooxygenase gene) enzyme of AOB and full denitrifiers (nitrous oxide reductase gene *nosZ*) in granules during reactor operation (Figure 7). Standard curves were made showing the relationship between the C_t threshold value and logarithm of the number of gene copies in the tested biomass. The curves were linear in the studied range of gene copy numbers, and the reactors determination coefficient was 0.99, 0.98 and 0.99 for the 16S rRNA, *amoA*, and *nosZ* genes, respectively. The number of gene copies in the above studies was referred to 5 ng DNA (unit of sample).

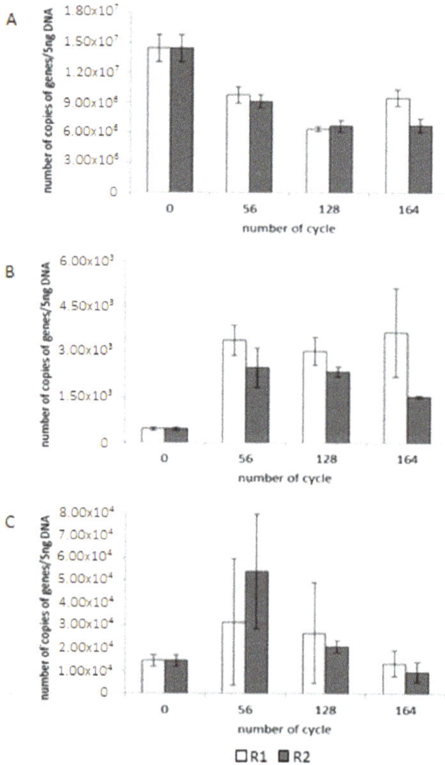

Figure 7. The absolute number of copies of (**A**) bacterial *16SrRNA*, (**B**) *amoA* and (**C**) *nosZ* genes in the reactors.

The abundance of the *16S rRNA* genes in the biomass of both reactors gradually decreased as the experiment was carried out. The largest number of 16S rRNA genes was recorded in the inoculum (1.4×10^7 copies/5 ng DNA). The number of copies of the 16S rRNA in mature granules at 164th cycle was at a level of 9.52×10^6/5 ng DNA and 6.75×10^6 copies/5 ng DNA in R_1 and R_2, respectively.

The abundance of bacteria with the ammonium monooxygenase (*amoA*) gene in the inoculum was 4.46×10^2copies/5 ng DNA. A rapid increase in the number of copies of the *amoA* gene was observed in R_1 at the beginning of the experiment, and during the experiment the number of copies remained in the range of 3.00–3.66×10^3copies/5 ng DNA. In R_2, the highest number of the investigated genes (2.49×10^3 copies/5 ng DNA) was recorded in 56th cycle of experiment. At the end of the experiment, the number of ammonium monooxygenase gene in biomass from R_1 was 2.4 times higher than in R_2. The copy number of the nitrous oxide reductase (*nosZ*) gene in DNA isolated from the inoculum was 1.48×10^4 copies/5 ng DNA. In R_1, the number of compiled denitrifying bacteria increased in the 56th cycle of reactor operation to 3.17×10^4 copies/5 ng DNA which could improve the *nosZ* gene expression and lead to higher emission of N_2 [46]. After that, this value slowly decreased to 1.35×10^4 copies/5 ng DNA (164th cycle). In R_2, the number of copies of the *nosZ* gene increased to 5.41×10^4 copies/5 ng DNA in the 56th cycle of the experiment thus the number of denitrification bacteria was 41% higher in the R_2 than in R_1. After this time, a reverse tendency was observed and at the end of the experiment the copy number of the investigated gene in biomass was about 1.4-time higher in R_1 than R_2.

Thwaites et al. [50], observed that the highest number of nitrifiers was recorded in the reactor with constant aeration but the introduction of a 60-min anaerobic phase into the working cycle lowered the number of nitrifiers while the highest number of denitrifying bacteria (*nosZ* gene analysis) characterized biomass from a reactor with a short 20-min anaerobic phase in the cycle. The highest efficiency of nitrogen removal from wastewater was obtained in the reactor with constant aeration (89%), while the lowest in the reactor with the initial 60-min anaerobic phase (58%). In the presented study, a lower number of denitrification bacteria in R_2 was observed despite a higher denitrification efficiency. This may be because denitrification numbers were estimated based on the *nosZ* gene analysis. Abundance of complete denitrifiers possessing *nosZ* gene in biomass is favorable because ensures full conversion of nitrogen compounds to N_2 and reduction in greenhouse gas emissions such as NO and N_2O [51]. Complete denitrification can be carried out by bacteria from the genera *Dokdonella*, *Flavobacterium*, *Acinetobacter*, *Pseudomonas*, *Arcobacter*, and *Comamonas*, while microorganisms carrying out partial denitrification belong to genera *Diaphorobacter*, *Thauera* and *Zoogloea* [52]. In the present study, in the biomass treating wastewater from the meat industry bacteria from genera *Dokdonella* (*Dokdonella ginsengisoli*) and *Thauera* (*Thauera* sp.) predominated. Microorganisms from the genera *Pseudomonas*, *Burkholderia* (*Burkholderia sp.*), *Geobacter*, and *Rhizobium* were also found in the biomass from R_2. These microorganisms have *norB* gene in their genomes, which product is responsible for the conversion of nitric oxide to nitrous oxide and can support the denitrification process. An increase in the occurrence of *Paracoccus* sp. was also noted at the end of the experiment in both reactors. *Paracoccus* sp. uses methanol as the carbon source and can denitrify at a high rate (250 mg NO_3/g TSS h^{-1}) [53,54]. The high percentage of unclassified microorganisms in R_2 indicates that more variable microenvironments created in biomass by intermittent aeration favored growth of yet not classified bacteria and resulted in higher microbial biodiversity.

Bacterial consortia in aerobic granules were analyzed using NGS to draw conclusion about the relationships between their structure and aeration mode in the reactor. The total number of 118,579 *16S rRNA* sequence reads assigned to different operational taxonomic units was obtained by Illumina MiSeq platform (Appendix A Table A2).

To assess the within–sample complexity of microbial population, the Shannon-Wiener diversity index (H') was evaluated. The Shannon-Wiener species diversity index in inoculum was 3.67. At the final stage of the experiment the index was by about 8% higher in R_2 (3.67) compared to R_1 (3.39).

At a species level (Table 1), the most abundant in inoculum was *Dokdonella ginsengisoli* (19.6%), *Tetrasphaera* sp. (3.8%), *Lysobacter daejeonensis* (3.7%), *Terrimonas arctica* (3.3%) and *Candidatus Competibacter*

(3.2%). In R_1 at the end of the experiment, biomass was dominated by *Dokdonella ginsengisoli* (20.0%), *Amaricoccus tamworthensis* (10.9%), *Thauera* sp. (10.7%), *Candidatus Competibacter* sp. (5.1%) and *Arenimonas* sp. (4.1%). *Dokdonella ginsengisoli* (16.0%), *Candidatus Competibacter* (7.7%), *Arenimonas* sp. (2.5%) and *Thauera* sp. (2.4%) dominated also at the end of the experiment in R_2.

During the experiment disappearance or reduction of the percentage was observed for *Acidobacterium* sp., *Tetrasphaera* sp., *Terrimonas arctica*, *Roseiflexus* sp., *Ignavibacterium* sp., *Ottowia* sp., *Rhodoferax* sp., *Nannocystis* sp., *Tolumonas* sp., *Aquimonas* sp. and *Lysobacter daejeonensis*.

In the present research, large changes in the structure of microorganisms colonizing the inoculum and granular sludge were noted throughout the experiment. Weissbrodt et al. [30], observed that during the granulation of activated sludge the percentage of *Competibacter* sp. increased and bacteria of the *Xanthomonadaceae*, *Rhodospirillaceae*, and *Aminobacter* family appeared. According to the authors, occurrence of such bacteria as *Zoogloea* sp., *Xanthomonadaceae*, *Sphingomonadales*, and *Rhizobiales* promotes the formation of EPS that are responsible for biomass granulation. In the present study, *Xanthomonadaceae* (*Dokdonella ginsengisoli*, *Lysobacter daejeonensis*) were found to be responsible for the formation of EPS and denitrification in biomass. The presence of microorganisms from the above-mentioned families promotes granulation and affects the strength of the resulting granules.

Bacterial taxa unique for the biomass from R_1 were identified including e.g., *Pseudonocardia babensis*, *Amaricoccus tamworthensis*, *Paracoccus* sp., *Paracoccus yeei*, *Arenimonas* sp. Many environmental bacteria isolated, among others, from water and soil possess the ability to break down fats, oils and greases [55]. The presence of complex compounds such as proteins, lipids or polysaccharides in wastewater stimulates growth of microorganisms in the biomass that carry out the hydrolysis of complex organic compounds, e.g., *Sphingobacteriales*, *Chitinophagaceae*, or *Flavobacterium* sp., which were also observed in the presented studies. In the granules, *Burkholderia* sp., and *Pseudomonas* sp. have been also identified that can break down lipids to fatty acids [56,57]. In the structure of aerobic granules, lipids occur peripherally in EPS, thus, the fats in the influent may attach to the surface of the granules and impede the transfer of oxygen deep into its structure increasing the demand of oxygen for microorganisms [58].

The ordination diagram (Figure 8) shows that at the level of the species variable aeration conditions in the reactor cycle did not affect the composition of microbiota ($p = 0.18$, $F = 1.66$), while the number of reactor working cycles significantly affected it ($p < 0.01$, $F = 3.61$). There was a strong positive correlation between the percentage share in the biomass of *Gemmobacter* sp. ($R = 0.98$), *Flexibacter* sp. ($R = 0.96$), *Paracoccus yeei* ($R = 0.85$) and *Rhodobacter* sp. ($R = 0.97$) and the experimental cycle indicating that these taxa were important for efficient treatment of meat-processing wastewater. A negative correlation was observed, among others, for *Ignavibacterium* sp. ($R = -0.96$), *Roseiflexus* sp. ($R = -0.94$) and *Tolumonas* sp. ($R = -0.92$).

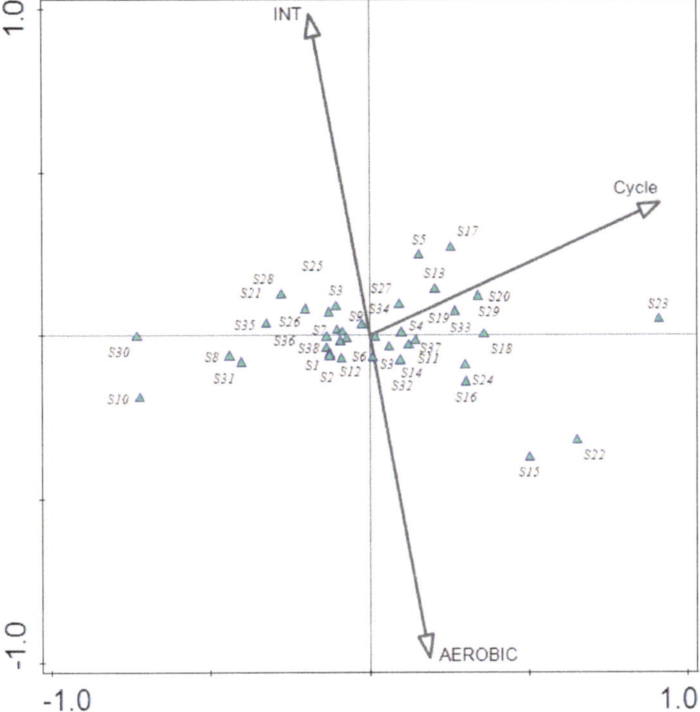

Figure 8. CCA of community of bacteria in the reactors. Environmental variables are represented by grey arrows (Cycle—Number of the reactor cycles, AEROBIC—Constant aeration in the reactor cycle, INT—intermittent aeration in the rector cycle). The taxa are named by letters and numbers: *Acidobacterium* sp. (S1), *Tetrasphaera* sp. (S2), *Pseudonocardia babensis* (S3), *Flexibacter* sp. (S4), *Flavobacterium* sp. (S5), *Terrimonas arctica* (S6), *Caldilinea* sp. (S7), *Roseiflexus* sp. (S8), *Trichococcus* sp. (S9), *Ignavibacterium* sp. (S10), *Nitrospira* sp. (S11), *Brevundimonas* sp. (S12), *Hyphomicrobium* sp. (S13), *Mesorhizobium* sp. (S14), *Amaricoccus macauensis* (S15), *Amaricoccus tamworthensis* (S16), *Gemmobacter* sp. (S17), *Paracoccus lutimaris* (S18), *Paracoccus* sp. (S19), *Paracoccus yeei* (S20), *Rhodobacter* sp. (S21), *Tetracoccus cechii* (S22), *Micavibrio* sp. (S23), *Burkholderia* sp. (S24), *Hydrogenophaga* sp. (S25), *Ottowia* sp. (S26), *Rhodoferax* sp. (S27), *Dechloromonas* sp. (S28), *Thauera* sp. (S29), *Nannocystis* sp. (S30), *Tolumonas* sp. (S31), *Steroidobacter* sp. (S32), *Pseudomonas* sp. (S33), *Candidatus Competibacter* (S34), *Aquimonas* sp. (S35), *Dokdonella ginsengisoli* (S36), *Arenimonas* sp. (S37), *Lysobacter daejeonensis* (S38).

4. Conclusions

The research allowed to draw the following conclusions:

- In an intermittently aerated reactor fed with high-TSS wastewater from meat industry higher N and P removal and a higher concentration of biomass was observed, but also a higher concentration of suspended solids in treated wastewater, smaller diameters of granules and deteriorated settling properties of biomass mainly as a result of the reduction of EPS content in biomass was noted;
- CCA analysis showed that microbial structure significantly changed in time; an increase in abundances of *Arenimonas* sp., *Thauera* sp., and *Dokdonella* sp. characterized mature granules in period of stable treatment of meat-processing wastewater;
- Constant aeration in the cycle increases the number of nitrifying and denitrifying genes in biomass (*amoA* and *nosZ* analysis) while intermittent aeration increases microbial diversity of granules;

- Constant aeration favored growth of microorganisms from the *Rhodanobacteraceae*, *Rhodobacteraceae* and *Xanthomonadaceae* families while in intermittently aerated reactors *Competibacteraceae* family was more abundant (7.8%) and appeared.

Author Contributions: Conceptualization, P.J., and A.C.-K.; methodology, P.J. and A.C.-K.; software, P.J.; validation, P.J.; formal analysis, P.J., and A.C.-K.; investigation, P.J., and P.S.; resources, P.J.; data curation, P.J., and A.C.-K.; writing—Original draft preparation, P.J.; writing—Review and editing, A.C.-K. and P.S.; visualization, P.J.; supervision, A.C.-K.; project administration, P.J.; funding acquisition, A.C.-K. All authors have read and agreed to the published version of the manuscript.

Funding: This research was funded by Minister of Science and Higher Education grant number 010/RID/2018/19 and Development Program at the University of Warmia and Mazury in Olsztyn grant number POWR.03.05. 00-00-Z310/17.

Acknowledgments: Project financially supported by Minister of Science and Higher Education in the range of the program entitled "Regional Initiative of Excellence" for the years 2019–2022, Project No. 010/RID/2018/19, amount of funding 12,000,000 PLN and was written as a result of the author's internship in Aalborg University, Denmark, co-financed by the European Union under the European Social Fund (Operational Program Knowledge Education Development), carried out in the project Development Program at the University of Warmia and Mazury in Olsztyn (POWR.03.05. 00-00-Z310/17).

Conflicts of Interest: The authors declare no conflict of interest.

Appendix A

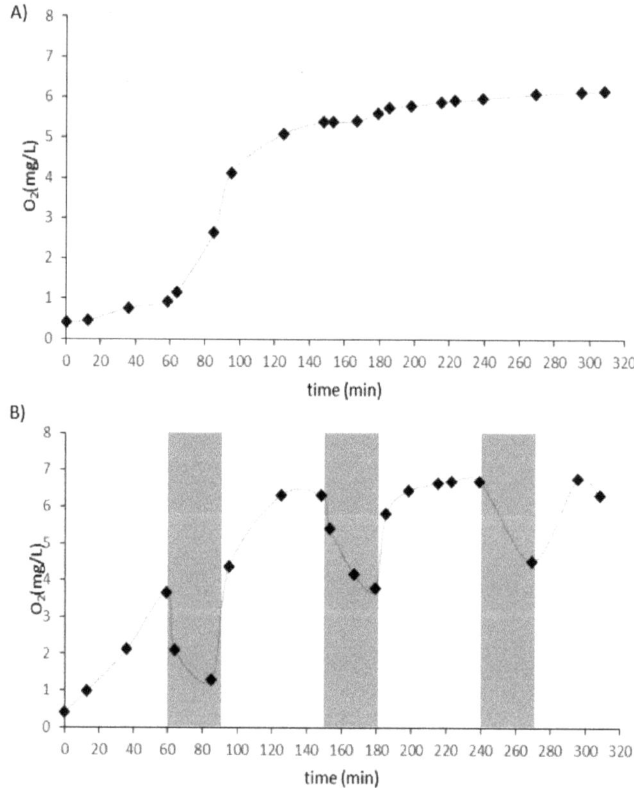

Figure A1. Changes of oxygen concentration in reactors cycle in (**A**) reactor R_1 and (**B**) R_2. Gray shading indicates phases without aeration.

Table A1. Primers and conditions applied in real-time PCR.

Gene, Primer	Primer Concentration	Size of Amplicon (bp)	Thermal Profile	Reference
amoA (amoA1Fa/amoA2R)	100 nM	~500	94 °C/15 s, 52 °C/45 s, 60 °C/45 s	[59]
Bacterial *16S rRNA* (519F/907R)	100 nM	~600	94 °C/15 s, 50 °C/40 s, 60 °C, 40 s	[60]
nosZ (NosZ-F/NosZ1622R)	200 nM	~500	94 °C/15 s, 60 °C/2.5 min	[61,62]

Table A2. Diversity indices from aerobic granular sludge sample in reactorss.

Reactor	Number of Cycle	Number of Sequences	Number of OTU	Shannon-Wiener Index (H′)
Inoculum	0	9675	994	3.67
R_1	56	20,360	873	3.22
	128	18,336	654	2.26
	164	26,594	901	3.39
R_2	56	21,699	947	3.02
	128	7237	740	3.59
	164	14,678	1031	3.67

References

1. Gerbens-Leenes, P.W.; Mekonnen, M.M.; Hoekstra, A.Y. The water footprint of poultry, pork and beef: A comparative study in different countries and production systems. *Water Resour. Ind.* **2013**, *1–2*, 25–36. [CrossRef]
2. The Food and Agriculture Organization. Available online: http://www.fao.org/faostat/en (accessed on 1 August 2020).
3. Pingali, P. Westernization of Asian diets and the transformation of food systems: Implications for research and policy. *Food Policy* **2007**, *32*, 281–298. [CrossRef]
4. Bustillo-Lecompte, C.F.; Mehrvar, M.; Quinones-Bolanos, E. Cost-effectiveness analysis of TOC removal from slaughterhouse wastewater using combined anaerobic-aerobic and UV/H_2O_2 processes. *J. Environ. Manag.* **2014**, *134*, 145–152. [CrossRef] [PubMed]
5. Kiepper, B.H.; Merka, W.C.; Fletcher, D.L. Proximate composition of poultry processing wastewater particulate matter from broiler slaughter plants. *Poult. Sci.* **2008**, *87*, 1633–1636. [CrossRef]
6. Cammarota, M.C.; Freire, D.G. A review on hydrolytic enzymes in the treatment of wastewater with high oil and grease content. *Bioresour. Technol.* **2006**, *97*, 2195–2210. [CrossRef] [PubMed]
7. Gao, D.; Liu, L.; Liang, H.; Wu, W.M. Aerobic granular sludge: Characterization, mechanism of granulation and application to wastewater treatment. *Crit. Rev. Biotechnol.* **2011**, *31*, 137–152. [CrossRef] [PubMed]
8. Liu, Y.; Yang, S.F.; Tay, J.H. Improved stability of aerobic granules by selecting slow-growing nitrifying bacteria. *J. Biotechnol.* **2004**, *108*, 161–169. [CrossRef] [PubMed]
9. Cydzik-Kwiatkowska, A.; Zielińska, M. Bacterial communities in full-scale wastewater treatment systems. *World J. Microbiol. Biotechnol.* **2016**, *32*, 66. [CrossRef] [PubMed]
10. Carrera, P.; Campo, R.; Méndez, R.; Di Bella, G.; Campos, J.L.; Mosquera-Corral, A.; Val del Río, A. Does the feeding strategy enhance the aerobic granular sludge stability treating saline effluents? *Chemosphere* **2019**, *226*, 865–873. [CrossRef] [PubMed]
11. Chen, F.Y.; Liu, Y.Q.; Tay, J.H.; Ning, P. Alternating anoxic/oxic condition combined with step-feeding mode for nitrogen removal in granular sequencing batch reactors (GSBRs). *Sep. Purif. Technol.* **2013**, *105*, 63–68. [CrossRef]
12. Cydzik-Kwiatkowska, A.; Wojnowska-Baryła, I. Nitrogen-converting communities in aerobic granules at different hydraulic retention times (HRTs) and operational modes. *World J. Microbiol. Biotechnol.* **2015**, *31*, 75–83. [CrossRef] [PubMed]

13. Liu, Y.; Kang, X.; Li, X.; Yuan, Y. Performance of aerobic granular sludge in a sequencing batch bioreactor for slaughterhouse wastewater treatment. *Bioresour. Technol.* **2015**, *190*, 487–491. [CrossRef] [PubMed]
14. Cassidy, D.P.; Belia, E. Nitrogen and phosphorus removal from an abattoir wastewater in a SBR with aerobic granular sludge. *Water Res.* **2005**, *39*, 4817–4823. [CrossRef] [PubMed]
15. Wingender, J.; Neu, T.R.; Flemming, H.C. (Eds.) *Microbial Extracellular Polymeric Substances: Characterization, Structures and Function*; Springer: Berlin/Heidelberg, Germany, 1999; pp. 1–18.
16. Sheng, G.P.; Yu, H.Q.; Li, X.Y. Extracellular polymeric substances (EPS) of microbial aggregates in biological wastewater treatment systems: A review. *Biotechnol. Adv.* **2010**, *28*, 882–894. [CrossRef] [PubMed]
17. Shin, H.S.; Kang, S.T.; Nam, S.Y. Effect of carbohydrate and protein in the EPS on sludge settling characteristics. *Water Sci. Technol.* **2001**, *43*, 193–196. [CrossRef]
18. Sponza, D.T. Investigation of extracellular polymer substances (EPS) and physicochemical properties of different activated sludge flocs under steady-state conditions. *Enzym. Microb. Technol.* **2003**, *32*, 375–385. [CrossRef]
19. Świątczak, P.; Cydzik-Kwiatkowska, A. Performance and microbial characteristics of biomass in a full-scale aerobic granular sludge wastewater treatment plant. *Environ. Sci. Pollut.* **2017**, *25*, 1655–1669. [CrossRef]
20. APHA (Ed.) *1992 Standard Methods for the Examination of Water and Wastewater*, 18th ed.; APHA: Washington, DC, USA, 1992.
21. Rusanowska, P.; Cydzik-Kwiatkowska, A.; Świątczak, P.; Wojnowska-Baryła, I. Changes in extracellular polymeric substances (EPS) content and composition in aerobic granule size-fractions during reactor cycles at different organic loads. *Bioresour. Technol.* **2019**, *272*, 188–193. [CrossRef]
22. Frølund, B.; Griebe, T.; Nielsen, P.H. Enzymatic activity in the activated-sludge floc matrix. *Appl. Microbiol. Biotechnol.* **1995**, *43*, 755–761. [CrossRef]
23. Świątczak, P.; Cydzik-Kwiatkowska, A.; Rusanowska, P. Microbiota of anaerobic digesters in a full-scale wastewater treatment plant. *Arch. Environ. Prot.* **2017**, *43*, 53–60. [CrossRef]
24. Hill, M.O. Diversity and evenness: A unifying notation and its consequences. *Ecology* **1973**, *54*, 427–432. [CrossRef]
25. Ter Braak, C.J.F.; Šmilauer, P. Canoco 5: Software for multivariate data exploration, testing and summarization. In *Biometrics*; Plant Research International: Wageningen, The Netherlands, 2012.
26. Mosquera-Corral, A.; De Kreuk, M.K.; Heijnen, J.J.; Van Loosdrecht, M.C. Effects of oxygen concentration on N-removal in an aerobic granular sludge reactor. *Water Res.* **2005**, *39*, 2676–2686. [CrossRef] [PubMed]
27. Tay, J.H.; Liu, Q.S.; Liu, Y. The effect of upflow air velocity on the structure of aerobic granules cultivated in a sequencing batch reactor. *Water Sci. Technol.* **2004**, *49*, 35–40. [CrossRef]
28. De Kreuk, M.K.; Pronk, M.; Van Loosdrecht, M.C. Formation of aerobic granules and conversion processes in an aerobic granular sludge reactor at moderate and low temperatures. *Water Res.* **2005**, *39*, 4476–4484. [CrossRef]
29. Luo, J.; Chen, H.; Han, X.; Sun, Y.; Yuan, Z.; Guo, J. Microbial community structure and biodiversity of size-fractionated granules in a partial nitritation–anammox process. *FEMS Microbiol. Ecol.* **2005**, *93*, fix021. [CrossRef]
30. Weissbrodt, D.G.; Neu, T.R.; Kuhlicke, U.; Rappaz, Y.; Holliger, C. Assessment of bacterial and structural dynamics in aerobic granular biofilms. *Front. Microbiol.* **2013**, *4*, 175. [CrossRef]
31. De Sousa Rollemberg, S.; De Oliveira, L.; Barros, A.; Melo, V.; Firmino, P.; Dos Santos, A. Effects of carbon source on the formation, stability, bioactivity and biodiversity of the aerobic granule sludge. *Bioresour. Technol.* **2019**, *278*, 195–204. [CrossRef]
32. Erşan, Y.Ç.; Erguder, T.H. The effect of seed sludge type on aerobic granulation via anoxic–aerobic operation. *Environ. Technol.* **2014**, *35*, 2928–2939. [CrossRef]
33. Ji, B.; Yang, K.; Zhu, L.; Jiang, Y.; Wang, H.; Zhou, J.; Zhang, H. Aerobic denitrification: A review of important advances of the last 30 years. *Biotechnol. Bioprocess Eng.* **2015**, *20*, 643–651. [CrossRef]
34. Cydzik-Kwiatkowska, A.; Rusanowska, P.; Zielińska, M.; Bernat, K.; Wojnowska-Baryła, I. Microbial structure and nitrogen compound conversions in aerobic granular sludge reactors with non-aeration phases and acetate pulse feeding. *Environ. Sci. Pollut. Res.* **2016**, *23*, 24857–24870. [CrossRef] [PubMed]
35. Griffiths, P.C.; Stratton, H.M.; Seviour, R.J. Environmental factors contributing to 385 the "G bacteria" population in full-scale EBPR plants. *Water Sci. Technol.* **2002**, *4–5*, 185–192. [CrossRef]

36. Zeng, R.J.; Yuan, Z.; Keller, J. Improved understanding of the interactions and complexities of biological nitrogen and phosphorus removal processes. *Rev. Environ. Sci. Biol.* **2004**, *3*, 265–272. [CrossRef]
37. Nielsen, P.H.; Mielczarek, A.T.; Kragelund, C. A conceptual ecosystem model of microbial communities in enhanced biological phosphorus removal plants. *Water Res.* **2010**, *44*, 5070–5088. [CrossRef] [PubMed]
38. Ge, H.; Batstone, D.J.; Keller, J. Biological phosphorus removal from abattoir wastewater at very short sludge ages mediated by novel PAO clade Comamonadaceae. *Water Res.* **2015**, *69*, 173–182. [CrossRef] [PubMed]
39. Mandel, A.; Zekker, I.; Jaagura, M.; Tenno, T. Enhancement of anoxic phosphorus uptake of denitrifying phosphorus removal process by biomass adaption. *Int. J. Environ. Sci. Technol.* **2019**, *16*, 5965–5978. [CrossRef]
40. Falvo, A.; Levantesi, C.; Rossetti, S.; Seviour, R.J.; Tandoi, V. Synthesis of intracellular storage polymers by *Amaricoccus kaplicensis*, a tetrad forming bacterium present in activated sludge. *J. Appl. Microbiol.* **2001**, *91*, 299–305. [CrossRef] [PubMed]
41. Saunders, A.M.; Oehmen, A.; Blackall, L.L. The effect of GAOs (glycogen accumulating organisms) on anaerobic carbon requirements in full-scale Australian EBPR (enhanced biologicalphosphorus removal) plants. *Water Sci. Technol.* **2003**, *47*, 37–43. [CrossRef] [PubMed]
42. Hynes, R.K.; Knowles, R. Production of nitrous oxide by Nitrosomonas europaea: Effects of acetylene, pH, and oxygen. *Can. J. Microbiol.* **1984**, *30*, 1397–1404. [CrossRef]
43. Wang, Z.P.; Liu, L.L.; Yao, H.; Cai, W.M. Effects of extracellular polymeric substances on aerobic granulation in sequencing batch reactors. *Chemosphere* **2006**, *63*, 1728–1735. [CrossRef]
44. Al-Halbouni, D.; Traber, J.; Lyko, S.; Wintgens, T.; Melin, T.; Tacke, D.; Hollender, J. Correlation of EPS content in activated sludge at different sludge retention times with membrane fouling phenomena. *Water Res.* **2008**, *42*, 1475–1488. [CrossRef]
45. McSwain, B.S.; Irvine, R.L.; Hausner, M.; Wilderer, P.A. Composition and distribution of extracellular polymeric substances in aerobic flocs and granular sludge. *Appl. Environ. Microbiol.* **2005**, *71*, 1051–1057. [CrossRef] [PubMed]
46. Adav, S.S.; Lee, D.J.; Tay, J.H. Extracellular polymeric substances and structural stability of aerobic granule. *Water Res.* **2008**, *42*, 1644–1650. [CrossRef] [PubMed]
47. Zhang, H.; Jia, Y.; Khanal, S.K.; Lu, H.; Fang, H.; Zhao, Q. Understanding the role of extracellular polymeric substances on ciprofloxacin adsorption in aerobic sludge, anaerobic sludge, and sulfate-reducing bacteria sludge systems. *Environ. Sci. Technol.* **2018**, *52*, 6476–6486. [CrossRef] [PubMed]
48. Wang, Y.X.; Lu, Z.X. Optimization of processing parameters for the mycelial growth and extracellular polysaccharide production by *Boletus* spp. ACCC 50328. *Process Biochem.* **2005**, *40*, 1043–1051. [CrossRef]
49. Chen, H.; Zhou, S.; Li, T. Impact of extracellular polymeric substances on the settlement ability of aerobic granular sludge. *Environ. Technol.* **2010**, *31*, 1601–1612. [CrossRef] [PubMed]
50. Thwaites, B.J.; Reeve, P.; Dinesh, N.; Short, M.D.; van den Akker, B. Comparison of an anaerobic feed and split anaerobic–aerobic feed on granular sludge development, performance and ecology. *Chemosphere* **2017**, *172*, 408–417. [CrossRef]
51. Jones, C.M.; Stres, B.; Rosenquist, M.; Hallin, S. Phylogenetic analysis of nitrite, nitric oxide, and nitrous oxide respiratory enzymes reveal a complex evolutionary history for denitrification. *Mol. Biol. Evol.* **2008**, *25*, 1955–1966. [CrossRef] [PubMed]
52. Pishgar, R.; Dominic, J.A.; Sheng, Z.; Tay, J.H. Denitrification performance and microbial versatility in response to different selection pressures. *Bioresour. Technol.* **2019**, *281*, 72–83. [CrossRef]
53. Srinandan, C.S.; Jadav, V.; Cecilia, D.; Nerurkar, A.S. Nutrients determine the spatial architecture of *Paracoccus* sp. biofilm. *Biofouling* **2010**, *26*, 449–459. [CrossRef]
54. Foglar, L.; Briški, F. Wastewater denitrification process—The influence of methanol and kinetic analysis. *Process Biochem.* **2003**, *39*, 95–103. [CrossRef]
55. Brooksbank, A.M.; Latchford, J.W.; Mudge, S.M. Degradation and modification of fats, oils and grease by commercial microbial supplements. *World J. Microbiol. Biotechnol.* **2007**, *23*, 977–985. [CrossRef]
56. Al-Khatib, M.A.; Alam, M.Z.; Shabana, H. Isolation of bacterial strain for biodegradation of fats, oil and grease. *Malays. J. Anal. Sci.* **2015**, *19*, 138–143.
57. Teixeira, P.D.; Silva, V.S.; Tenreiro, R. Integrated selection and identification of bacteria from polluted sites for biodegradation of lipids. *Int. Microbiol.* **2019**, *1*, 1–14. [CrossRef]
58. Wang, S.; Shi, W.; Tang, T.; Wang, Y.; Zhi, L.; Lv, J.; Li, J. Function of quorum sensing and cell signaling in the formation of aerobic granular sludge. *Rev. Environ. Sci. Bio/Technol.* **2017**, *16*, 1–13. [CrossRef]

59. Rotthauwe, J.H.; Witzel, K.P.; Liesack, W. The ammonia monooxygenase structural gene *amoA* as a functional marker: Molecular fine-scale analysis of natural ammonia-oxidizing populations. *Appl. Environ. Microbiol.* **1997**, *63*, 4704–4712. [CrossRef] [PubMed]
60. Lane, D.J. 16S/23S rRNA sequencing. In *Nucleic Acid Techniques in Bacterial Systematics*; Stackebrandt, E., Goodfellow, M., Eds.; Wiley: New York, NY, USA, 1991; pp. 205–248.
61. Kloos, K.; Mergel, A.; Rösch, C.; Bothe, H. Denitrification within the genus Azospirillum and other associative bacteria. *Aust. J. Plant Physiol.* **2001**, *28*, 991–998. [CrossRef]
62. Throbäck, I.N.; Enwall, K.; Jarvis, Ä.; Hallin, S. Reassessing PCR primers targeting *nirS*, *nirK* and *nosZ* genes for community surveys of denitrifying bacteria with DGGE. *FEMS Microbiol. Ecol.* **2004**, *49*, 401–417. [CrossRef]

Review

Microbial Activity in Subterranean Ecosystems: Recent Advances

Tamara Martin-Pozas [1,†], Jose Luis Gonzalez-Pimentel [2,†], Valme Jurado [3], Soledad Cuezva [4], Irene Dominguez-Moñino [3], Angel Fernandez-Cortes [5], Juan Carlos Cañaveras [6], Sergio Sanchez-Moral [1] and Cesareo Saiz-Jimenez [3,*]

1 Museo Nacional de Ciencias Naturales (MNCN-CSIC), 28006 Madrid, Spain; tmpozas@mncn.csic.es (T.M.-P.); ssmilk@mncn.csic.es (S.S.-M.)
2 Laboratorio HERCULES, Universidade de Evora, 7000-809 Evora, Portugal; pimentel@irnas.csic.es
3 Instituto de Recursos Naturales y Agrobiologia (IRNAS-CSIC), 41012 Sevilla, Spain; vjurado@irnase.csic.es (V.J.); idominguez@irnase.csic.es (I.D.-M.)
4 Departamento de Geología, Geografía y Ciencias Ambientales, Universidad de Alcala de Henares, 28805 Madrid, Spain; soledadcuezva@uah.es
5 Departmento de Biología y Geología, Universidad de Almeria, 04120 Almeria, Spain; acortes@ual.es
6 Departamento de Ciencias de la Tierra, Universidad de Alicante, 03080 Alicante, Spain; jc.canaveras@ua.es
* Correspondence: saiz@irnase.csic.es
† These authors contributed equally to this study.

Received: 19 October 2020; Accepted: 13 November 2020; Published: 17 November 2020

Abstract: Of the several critical challenges present in environmental microbiology today, one is the assessment of the contribution of microorganisms in the carbon cycle in the Earth-climate system. Karstic subterranean ecosystems have been overlooked until recently. Covering up to 25% of the land surface and acting as a rapid CH_4 sink and alternately as a CO_2 source or sink, karstic subterranean ecosystems play a decisive role in the carbon cycle in terms of their contribution to the global balance of greenhouse gases. Recent data indicate that microbiota must play a significant ecological role in the biogeochemical processes that control the composition of the subterranean atmosphere, as well as in the availability of nutrients for the ecosystem. Nevertheless, there are still essential gaps in our knowledge concerning the budgets of greenhouse gases at the ecosystem scale and the possible feedback mechanisms between environmental-microclimatic conditions and the rates and type of activity of microbial communities in subterranean ecosystems. Another challenge is searching for bioactive compounds (antibiotics) used for treating human diseases. At present, there is a global health emergency and a strong need for novel biomolecules. In recent decades, great research efforts have been made to extract antibiotics from marine organisms. More recently, caves have been receiving considerable attention in search of novel antibiotics. Cave methanotrophic and heterotrophic bacteria are producers of bioactive compounds and may be potential sources of metabolites with antibacterial, antifungal or anticancer activities of interest in pharmacological and medical research, as well as enzymes with a further biotechnological use. Here we also show that bacteria isolated from mines, a still unexplored niche for scientists in search of novel compounds, can be a source of novel secondary metabolites.

Keywords: karst; methane; carbon dioxide; greenhouse gases; methanotrophy; cave bacteria; bioactive compounds

1. Introduction

Karst is the term used to describe terrains underlain by soluble rock and characterized by the occurrence of caves, sinkholes, sinking streams, and an assortment of other landforms carved on the bedrock. Shallow karst ecosystems cover up to 25% of the Earth's land surface [1] and differ from the surface environments because of their limited energy and available nutrients.

Caves, in general, are characterized by a constant temperature, humidity, and high carbon dioxide (CO_2) concentration the year round, as well as absence of light and scarcity of nutrients. Microorganisms occupy all the niches of the biosphere, including the subsurface, as a part of the critical zone, the heterogeneous near surface environment in which complex interactions involve rock, soil, water, air, and living organisms [2].

Earth's subsurface contains an active microbiota colonizing rock surfaces. In this environment, microorganisms are forced to adapt their metabolism for surviving in extreme conditions, and the low input of carbon, nitrogen and phosphorus as well as the chemical composition of the rock has a direct impact on the community diversity. In fact, one of the main reservoirs of microbial life, even at great depths, where life is not dependent on solar energy and photosynthesis for its primary energy supply is the terrestrial subsurface [3].

The colonization of substrates in caves is not homogeneous. Microorganisms colonize speleothems, host rock, detrital sediments, and/or speleosols with different compositions (clays, carbonate minerals, etc.) and/or textures (crystal habit, grain size, permeability, etc.). Microbial colonization is ultimately a complex and dynamic process that is determined and controlled by physicochemical properties (temperature, pH, redox potential, salinity) and biochemical factors (bioreceptivity, nutrient or electron acceptor availability, carbon, nitrogen and phosphorous concentrations, etc.) [4]. Therefore, the collective metabolic processes of microorganisms are decisive in the biogeochemical cycles of the biosphere: C and N fixation, CH_4 metabolism, S oxide reduction, etc.

It is well-known that dissolution and precipitation of carbonates are the main processes involved in the mobilization of carbon in subterranean environments. Cave microorganisms are able to induce the precipitation of carbonates, via biomineralization processes [5] and also dissolution processes due to the excretion of acids [6]. There is a wide array of literature on the study of bio-induced mineral formations in subterranean environments [7] and on the microbial–rock interaction related to the CO_2 uptake or release processes [8]. In this context, previous studies have confirmed that *Actinobacteria* biofilms developing on cave walls promote uptake of CO_2, dissolve the rock, and produce calcite crystals in periods of lower humidity and/or CO_2 [8]. However, the interactions of microbes with the air–water–rock interfaces in subterranean ecosystems and the biological mechanisms by which microorganisms adjust to new environments or changes in their current environment are poorly understood.

Low energy subsurface environments are uniquely positioned for examining minimum energetic requirements and adaptations for chemolithotrophic life and become a suitable environment to study the origins of life on Earth and may also serve as analogs to explore subsurface life in extraterrestrial bodies [9]. Furthermore, the microbiota from shallow subsurface environments (karst cavities, lava tubes) are becoming a target of increasing interest in different research fields, including biodiversity [10], mineral formation and dissolution [7], cultural and natural heritage conservation [11], and paleoclimatology [12]. In addition, other important uses of microorganisms are the production of bioactive compounds valuable for medicine and enzymes for bioremediation [13].

The extensive literature about microbial diversity and activity of cave microorganisms has been reviewed by many authors. The books "Microbial Life of Cave Systems" by Engel [14] and "Cave Ecology" by Moldovan et al. [15] are a rich source of information. In addition, other book chapters and review articles are relevant [16–21]. Because of the comprehensive scope of the literature on this topic, for this review we have selected two emerging research topics representing recent advances in environmental microbiology: (1) the control of greenhouse gas fluxes by cave microorganisms, and (2) the search of antibiotics produced by subsurface bacteria.

2. The Control of Greenhouse Gas Fluxes by Cave Microorganisms

Global changes in the Earth's climate and its relationship to the increasing concentration of greenhouse gases (GHGs) in the atmosphere has received special attention since the last quarter of the 20th century. Etiope and Klusman [22] reported that the major sources for atmospheric methane (CH_4) budget derive from the natural processes in the biosphere (modern microbial activity) and from fossil, radiocarbon-free CH_4 emission, estimated at approximately 20% of atmospheric CH_4, which is due to and mediated by anthropogenic activity. However, this estimation is higher than the estimates from statistical data of CH_4 emission from fossil fuel and related anthropogenic sources. For these authors, geologic sources are more than enough to provide the amount of CH_4 required to account for the suspected missing source of fossil CH_4. In addition, Etiope and Lollar [23] distinguished between biotic and abiotic CH_4, the latter produced in magmatic processes (volcanoes and high-temperature active hydrothermal vents) and postmagmatic processes at lower temperatures (gas–water–rock interactions).

A better understanding of the carbon cycle in the Earth-climate system is nowadays a crucial knowledge gap. The main research efforts are focused on identifying and characterizing all possible sources, reservoirs, and sinks of GHGs, mainly CO_2 and CH_4, in order to more accurately calculate the budgets, especially in the carbon cycle [24]. This issue is critical to understand the effects of changes in the carbon cycle on Earth's climate, and to assess the level of effort required in order to adapt and mitigate climate change.

The interactions between geological, microbiological, and chemical processes are responsible for the physical-chemical properties of the atmosphere and especially for changes in its composition. Caves and other shallow vadose environments are populated by methanotrophic microorganisms and thus represent a CH_4 sink. This subterranean CH_4 sink is largely overlooked in the scientific literature. Understanding how cave microbiomes influence the systems in which they inhabit is proving to be an exceptional research challenge [25].

Methane is consumed from the atmosphere by methanotrophs in forests, grasslands, paddy, and other unsaturated soils, which represent the major terrestrial sinks. Environmental CH_4 oxidation by bacteria is mainly carried out by *Gammaproteobacteria*, *Alphaproteobacteria*, and *Verrucomicrobia* [26], though there is also recent evidence for methanotrophy in *Rokubacteria* [27].

The presence of methanotrophic bacteria in caves has been widely studied in Movile Cave, Romania, by using isolation techniques, $^{13}CH_4$-labelling, and ^{13}C-DNA analysis, and the significant importance to the ecosystem development and primary productivity has been remarked upon [28–31]. Evidence of the occurrence of methanotrophs has also been found in other caves [10,32,33]. However, in these studies the microorganisms were not related with the sink of GHGs in caves.

Specific studies, both on the environmental-driven controls on microbial activity and, in turn, on the microbial role in composition changes of natural subterranean ecosystems, constitute a new research area of the highest potential with a pool of questions to solve. The starting hypothesis was that the subterranean microbiome plays a significant ecological role in the biogeochemical processes controlling the composition of the underground atmosphere, as well as in the availability of nutrients for the rest of the ecosystem's biota.

Fernandez-Cortes et al. [34] evidenced for the first time that cave ecosystems act as effective natural sinks of atmospheric CH_4 on seasonal and daily scales and this phenomenon may thus be relevant on a global scale in terms of its contribution to the global balance of GHGs. The potential methanotrophy in four Spanish caves was assessed by tracking the presence of methane-oxidizing bacteria using the particulate methane monooxygenase gene *pmoA*, which is a phylogenetic marker for identifying methanotroph-specific DNA sequences in the environment [35]. The study revealed the presence of the proteobacteria *Methylocapsa aurea*, *Methylomicrobium album*, *Methylococcus capsulatus*, and methanotrophs of the K1-1 and K3-16 groups in samples from Altamira, Sidron, and Ojo Guareña caves, mainly in locations where CH_4 usually reaches concentrations near to the atmospheric background levels. These soil bacteria oxidize the atmospheric CH_4 [36].

However, the analyses did not detect methanotrophs in remote subterranean locations or poorly ventilated caves, such as Castañar de Ibor Cave, where CH_4 is absent or present in minimal concentrations (below the accuracy threshold) throughout the year. Fernandez-Cortes et al. [34] suggested that complete consumption of CH_4 was favored in the subsurface atmosphere under near vapor-saturation conditions without significant intervention of methanotrophic bacteria. This led to the assumption that CH_4 oxidation was induced by ions and •OH generated by the radioactive decay of radon (^{222}Rn). In fact, one of the important •OH sources in cave air may be from radioactive ^{222}Rn decay [34]. However, further research verified that the mechanism of CH_4 consumption was seasonally changing and methane-oxidizing bacteria were primarily responsible for the widespread observations of CH_4 depletion in subterranean environments, discarding any evidence of radiolysis contribution [37–39].

Schimmelmann et al. [37] tested, in controlled laboratory experiments, whether radiolysis could rapidly oxidize CH_4 in sealed air with different relative humidity and elevated levels of radiation from Rn isotopes. No evidence of CH_4 oxidation by radiolysis was found. On the contrary, a rapid loss of CH_4 was found when moist soil in the absence of Rn was added to the container. This was consistent with the presence of methane-oxidizing bacteria, which were responsible for the widespread observations of CH_4 depletion in subterranean environments.

Since the pioneering work of Fernandez-Cortes et al. [34], a few authors, based on studies in caves from Australia, the USA, Vietnam, and Spain, additionally supported CH_4 oxidation by methanotrophic bacteria [38–42].

Webster et al. [38] reported that the concentrations and stable isotopic compositions of CH_4, CO_2, and Rn in cave air overlapped and diverged from those of the atmosphere, as the majority of cave air samples were depleted in CH_4 and enriched in CO_2 relative to the local atmosphere. These differences indicate that atmospheric and internal cave processes influenced the composition of cave air. Therefore, the authors, on the basis of CH_4 concentrations, $\delta^{13}C_{CH4}$, and $\delta^{2}H_{CH4}$ values measured in 33 caves in the USA and three caves in New Zealand, suggested that microbial methanotrophy within caves is the primary CH_4 consumption mechanism. Furthermore, the stable isotopic composition of CH_4 in the studied caves suggested that, in addition to atmospheric CH_4, at least two additional CH_4 sources were present in some caves: CH_4 produced from acetate fermentation, and from CO_2 reduction, processes occurring over a wide scale in the environment.

Lennon et al. [39] also proposed that biological processes, largely oxidation by methanotrophic bacteria, cause a depletion of CH_4 in caves. They conducted a field mesocosm experiment to test whether or not microbial methanotrophy has the potential to act as a daily sink for CH_4 in two fairly well-ventilated Vietnamese caves with low Rn concentrations (75–115 Bq/m^3), temperatures of 19–21 °C, and relative humidity ranging between 85 and 95%, depending on the airflow and location within the cave. The data suggested that biological processes have the potential to deplete atmospheric levels of CH_4 (~2 ppmv) via methanotrophy on a daily basis, as an 87% reduction in CH_4 concentrations was observed.

It appears that CH_4 depletion is a seasonal phenomenon, as reported by several authors. Fernandez-Cortes et al. [34] found significant seasonal and even daily variations in the gas composition of cave air, which involves the exchange of large amounts of other GHGs, in addition to CO_2(g), with the lower troposphere. Waring et al. [40] performed a continuous 3-year record of CH_4 and other trace gases in an Australian cave and found a seasonal cycle of extreme CH_4 depletion, from ambient ~1775 ppb to near zero during summer and to ~800 ppb in winter.

Ojeda et al. [41] found methanotrophic bacteria from the families *Methylococcaceae* (*Gammaproteobacteria*) and *Methylocystaceae* (*Alphaproteobacteria*) in 67% of the samples collected in Nerja Cave, Spain. In a recent innovative research, Cuezva et al. [42] confirmed that microbial action in caves plays a crucial role in the processes of production, consumption and storage of GHGs (CO_2 and CH_4) and largely determines the strong variations of these major GHGs in natural underground ecosystems. This study was developed in three Spanish caves (Pindal, Castañar de Ibor and La Garma) as a first approach to systematically characterize the role of cave sediments in the production and transport of CO_2 and CH_4 in the subterranean environment.

Monitoring and sampling for more than two years in La Garma Cave showed that during the stages with greater ventilation, air circulates daily and there is a continual contribution of external air to the cave, which has lower CO_2 content and CH_4 levels close to the atmospheric background. Therefore, CH_4 depletion rises with slight changes in CO_2. Conversely, in stages with a low ventilation rate, CO_2 reaches high concentrations in the cave because air exchange with the external atmosphere is negligible. Thus, the removed CH_4 is not rapidly replenished. As a result, CH_4 depletion rate tends to become negligible as the CO_2 content of cave air rises (Figure 1a).

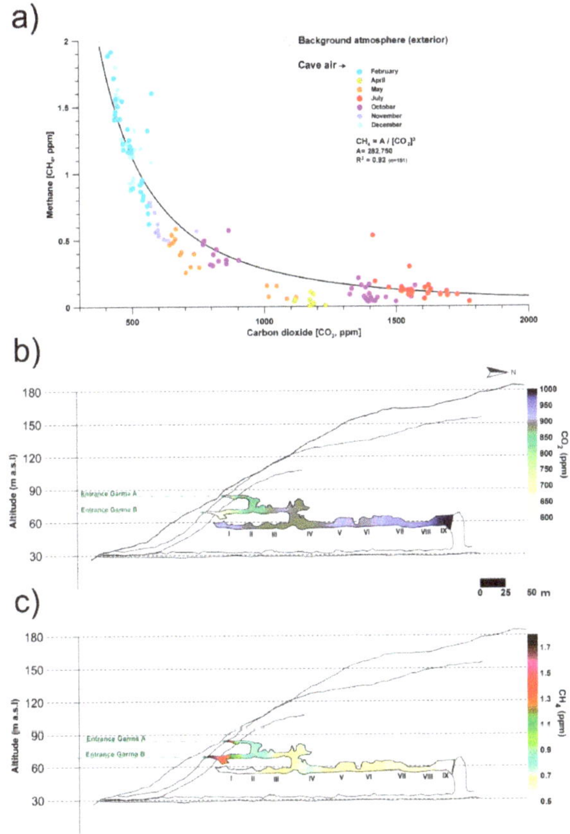

Figure 1. (**a**) Monthly co-variations in the concentrations of carbon dioxide (CO_2) and methane (CH_4) in La Garma Cave (Cantabria, northern Spain), a dynamically ventilated cave. (**b**) Spatial distribution of average concentrations of CO_2. (**c**) Spatial distribution of average concentrations of CH_4. Data from October 2014 to July 2017.

Figure 1b,c shows the spatial distribution of air CO_2 and CH_4 concentrations, respectively, in the air from La Garma Cave. Data for each contour map correspond to mean values from a set of bimonthly spot air samplings, conducted from October 2014 to July 2017, in a pre-established network of 11 points covering up to three levels of cave passages along an altitude gradient.

The average CO_2 and CH_4 concentrations of cave air were 894 and 0.65 ppm, respectively. Both GHGs depend on the rate of cave air exchange with the local atmosphere, which is controlled by climate-driven processes (primarily advection), and it is a very good indicator of the levels of matter and energy exchange with the exterior, showing the isolated areas and those with a prevailing connection with the exterior. Thus, a remarkable spatial pattern is distinguished; the highest average values of CO_2 concentration and the lowest CH_4 were found in the sectors of the lower gallery furthest from the main cave entrances (Garma A and Garma B, Figure 1b,c). Therefore, these cave maps with the contoured CO_2 and CH_4 levels reveal the importance of cave morphology in complex subterranean systems which control the gaseous composition of cave air, particularly in terms of gas variations due to the occurrence of elevation changes, multiple entrances or presence of dead-end passages. In the case of CH_4, its average concentration decreased drastically below 0.7 ppm from the connection of the intermediate gallery with the lower gallery and was practically null (<0.5 ppm) in the most distant sectors of the cave entrances (Figure 1c). This CH_4 pattern results from a decreasing percentage of mixing with the exterior and, consequently, a more effective methanotrophic activity of bacterial origin.

Cuezva et al. [42] are developing seasonal campaigns for CH_4 and CO_2 daily fluxes with continuous monitoring by a closed chamber-based gas exchange system (LI-COR Automated Soil Gas Flux System), in conjunction with a compatible Gasmet Fourier Transform Infrared (FTIR) gas analyzer and combined with $\delta^{13}C$ geochemical tracing by cavity ring-down spectroscopy (CRDS) to understand the underlying mechanisms in cave sediments. Moreover, an autonomous piece of equipment monitored the main microenvironmental parameters of the local subsurface-soil-atmosphere system. Preliminary results showed net CO_2 emissions from cave sediments resulting from respiration by chemolithotrophic microorganisms. The results also revealed simultaneous net CH_4 uptake from cave sediments on a daily scale, with no significant level of variations along the day (Figure 2). Anaerobic oxidation of CH_4 coupled to nitrite reduction is produced by members of the phylum *Rokubacteria*. These bacteria have also been found in Pindal Cave [42] and in an Alpine cave [43].

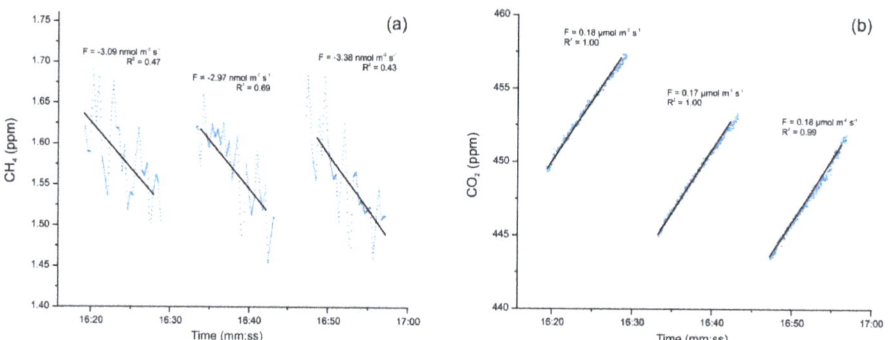

Figure 2. (**a**) Detail of CH_4 uptake fluxes with an average of -3.15 nmol m^{-2} s^{-1} and (**b**) simultaneous CO_2 emission fluxes uptake fluxes with an average of +0.17 μmol m^{-2} s^{-1}, monitored on 17 December 2019 directly above sediments inside Pindal Cave (Asturias, Spain). The value of the diffusive flux (F) and the corresponding exponential adjustment (R^2) of each measurement are indicated.

Other studies combining the depletion of CH_4 with other GHGs (N_2O and NO_2) were carried out in Vapor Cave, Southeast Spain. This is a hypogenic cave formed by the upwelling of hydrothermal CO_2-rich fluids in which anomalous concentrations of nitrogen oxides can be found [44]. The cave is characterized by a combination of rising warm air with large CO_2 outgassing and highly diluted CH_4 of endogenous origin. Additionally, extreme environmental conditions were observed, such as high air temperatures (38–43 °C) and 100% relative humidity, hypoxic conditions (17% O_2), CO_2 concentrations that exceed 1%, ^{222}Rn activity with values above 50 kBq/m^3, and a vertical thermal gradient of 3.2 °C/100 m [45]. These conditions, associated with the combined effects of tectonic activity and hydrothermalism, make this cave a remarkable site for the study of uncommon or extremophilic microbial communities. In Vapor Cave, the depletion of CH_4 was quantified to account for more than 60% removal of the deep endogenous component of this gas [45].

Martin-Pozas et al. [44] collected different cave air and sediment samples from −2 to −80 m in Vapor Cave. The analyses were conducted by taking advantage of technological advances in high-precision field-deployable CRDS and FTIR spectrometers, which allowed to measure target tracer gases (NO_2, N_2O, CH_4, and CO_2) and $\delta^{13}C$ of both carbon-GHGs in situ. The $\delta^{13}C_{CO2}$ data (−4.5 to −7.5‰) suggested a mantle-rooted CO_2 likely generated by the thermal decarbonation of underlying marine carbonates, combined with degassing from CO_2-rich groundwater. CH_4 molar fractions and their δD (−77 to −48‰) and $\delta^{13}C$ values (−52 to −30‰) indicated that the CH_4 reaching Vapor Cave is the remnant of a larger and deep-sourced CH_4, which was likely generated by the microbial reduction in carbonates. This CH_4 was affected by a postgenetic depletion during its migration through the cave environment as a component of the rising warm air.

CH_4 concentrations and $\delta^{13}CH_4$ varied with depth. At −80 m, higher concentrations were found but above −30 m depth lower CH_4 concentrations were found and heavier $\delta^{13}C$ values were found near the cave entrance. This was consistent with a methane oxidation mediated by microorganisms and in fact, next generation sequencing (NGS) analysis of sediments showed a relative abundance of *Candidatus* Methylomirabilis 4 to 5 times higher in the deepest sample (−80 m) with respect to −30 and −15 m. *Candidatus* Methylomirabilis oxyfera (*Rokubacteria*) is an anaerobic denitrifying methanotroph [46]. It must be noticed that Isobe et al. [47] found that members of the uncultivated candidate phylum *Rokubacteria* responded positively to elevate CO_2 concentrations.

In a similar way, Cappelleti et al. [48] studied an area of agricultural soils in Italy with anomalously high temperatures (up to ≅ 50 °C) and found emissions of biogenic CO_2 linked to CH_4 oxidation at a depth of 0.7 m from the surface. A strong biological methane-oxidizing activity in these soils was found and an examination of the *pmoA* clone libraries revealed the large biodiversity of methanotrophs including *Methylomonas*, *Methylococcus*, *Methylocystis*, and *Methylocaldum*.

Regarding the nitrogen gases, Martin-Pozas et al. [44] stated that the analysis of the ecological functions and metabolism of the microbiota from cave sediments suggested that N_2O is mainly produced in the deepest areas of Vapor Cave (below −15 m depth). In these areas, high CO_2 concentrations and low O_2 levels within the sediments determine a prevailing hypoxic and acidic environment that promotes the release of nitrite, nitric oxide, and hydroxylamine as products of the metabolism of ammonia-oxidizing archaea and nitrate reduction. In fact, at −15 m depth, the archaeal communities were dominated by the class *Nitrososphaeria* (69.0% of the total *Archaea*), with a majority of uncultured members and only two identified genera, *Nitrososphaera* and *Nitrosotenuis*. This is consistent with the abundant occurrence of these *Archaea* in deep sediments and better survival under conditions of low dissolved oxygen.

To summarize, considerable advances have been reached in recent years regarding processes of production, consumption, and storage of greenhouse gases (CO_2, CH_4 and N_xO_x) by cave microorganisms in subterranean vadose ecosystems. Recent and current research has shown that cave *Actinobacteria* are active agents in the fixation of CO_2, capturing CO_2 from air and forming calcium carbonate polymorphs [8]. In particular, direct CO_2 flux measurements in areas heavily colonized by bacteria indicate that they were promoting the uptake of this gas. Subterranean environments act as sinks or net sources of soil-derived carbon dioxide (CO_2) on annual and even daily scales, reaching up to ten times higher than the mean atmospheric CO_2 content, which involves the exchange of large amounts of CO_2(g) with the lower troposphere and its role as a depot and/or emitter. In a very recent in-situ experimental work (Pindal Cave, Spain) with a closed chamber-based gas exchange system—research in progress—we have verified negative CH_4 fluxes (uptake) from microbial communities, simultaneously linked to positive CO_2 fluxes (emission) directly related to microbial methanotrophy. The most recent data from direct measurements of gas exchange fluxes indicate that both gases are inextricably linked in these microbial-induced processes.

3. The Search of Antibiotics Produced by Subsurface Bacteria

Methanotrophic bacteria not only perform consumption of CH_4 in caves, but also have other capabilities. Methanotrophs produce methanobactins, bioactive compounds which have a high affinity for copper, but can bind additional metal ions, suggesting that these compounds might play a role in the protection against toxicity of metal ions other than copper [49]. Methanobactins exhibited antibiotic activity against Gram-positive bacteria and have been investigated as a treatment for Wilson disease, a human disorder involving toxic copper accumulation. For more detailed information, the reader is referred to the review of Kenney and Rosenzweig [49] (and references therein). Other uses of methanotrophs rely on their application and potential value in bioremediation, namely methane removal from landfills and coal mines, as well as biodegradation of toxic compounds [50,51].

In the biosphere, methanotrophs share niches with other bacterial groups, namely *Proteobacteria* and *Actinobacteria* [43,44]. In fact, several reports have shown the interactions between methanotrophs and heterotrophs and clear beneficial associations promoting growth were observed among them [52–54]. Most of these heterotrophs are well-known producers of bioactive compounds (antibiotics).

3.1. Why Is There a Need of New Antibiotics?

According to the World Health Organization (WHO) [55], there is a serious lack of new antibiotics to combat the growing threat of antimicrobial resistance of pathogenic bacteria as well as an urgent need for more investment in research and development to fight against antibiotic-resistant infections. De Kraker et al. [56] predicted more than ten million deaths of people infected with the antibiotic-resistant bacteria worldwide within the next 30 years if urgent measures are not taken now.

In 2017, the WHO [57] reported a list of 12 bacteria for which novel antibiotics were urgently required. The list comprises three categories: critical, high and medium priority, according to the urgency of need for new antibiotics. In critical priority were included *Acinetobacter baumannii*, *Pseudomonas aeruginosa*, and *Enterobacteriaceae*. High priority was assigned to *Enterococcus faecium*, *Staphylococcus aureus*, *Helicobacter pylori*, *Campylobacter* spp., *Salmonellae*, and *Neisseria gonorrhoeae*, and finally the category of medium priority encompasses *Streptococcus pneumoniae*, *Haemophilus influenzae*, and *Shigella* spp. It is remarkable that only three out of the twelve bacteria from the list are Gram-positive, which denotes the interest in having antibiotics active against Gram-negative bacteria.

The clinical pipeline for new antibiotics currently includes only a small number of novel compounds in development. In the past 20 years, only two new antibiotic classes, both only active against Gram-positive bacteria, have received global regulatory approval by the US Food and Drug Administration and European Medicines Agency [58]. In the same time period, no new antibiotics against Gram-negative bacteria have been approved, and the quinolones, discovered in 1962, represent the last novel drug class identified to be active against Gram-negative bacteria. In addition, only two completely new drugs for multidrug-resistant tuberculosis treatment have reached the market in over 70 years [58].

Currently, the clinical pipeline is dominated by improvements of existing products with similar chemical structures, but some level of cross-resistance and fast adaptation of target bacteria can be expected. Resistance to one specific antibiotic agent can lead to resistance to a whole related class, through exchange of genetic material between different bacteria, and can affect antibiotic treatment of a wide range of infections and diseases [59]. Ideally, research should be focused on entirely new classes of compounds, targets, and modes of action to avoid cross-resistance to existing antibiotics [60].

3.2. Antibiotics from Subsurface Bacteria

About two-thirds of all known antibiotics were produced by *Actinobacteria*, particularly by species of the genus *Streptomyces* [61,62]. Historically, the decades from the 1930s to 1980s represented the golden age of antibiotic discovery. The starting point was the exhaustive screening on soil *Actinobacteria* carried out by Waksman and collaborators, which led to the discovery of actinomycin and streptomycin [63,64].

Further screenings of soil bacteria produced thousands of bioactive compounds, such as the well-known chloramphenicol, tetracycline, erythromycin, vancomycin, gentamicin, etc. However, in subsequent searches the same or similar compounds already known were recurrently obtained from soil bacteria. Even different species produced the same compounds [65]. Therefore, the investigations adopted different strategies.

A few authors stated that new bioactive compounds can be found in *Actinobacteria* that have been previously studied to the extent that re-examination of known microorganisms, already in storage, should provide novel compounds [66,67]. In this context, Takahashi and Nakashima [67] tested lyophilized actinobacterial strains from the Kitasato University Microbial Library, Japan, most of them more than 35 years old, and found that 330 strains were producers of useful bioactive compounds. In this work, a strain of *Streptomyces griseus*, isolated from a soil sample in 1971, and producing streptomycin, was cultivated in four different media and revealed to yield two new N-containing compounds, iminimycin A and B, both with antimicrobial activity against Gram-positive and Gram-negative bacteria. This finding was possible due to the methodology adopted, a physico-chemical screening of culture broths involving the use of a reagent to identify nitrogen-containing metabolites and a routine analysis by liquid chromatography–mass spectrometry, liquid chromatography–UV detection, and polarity and their comparison with existing databases.

It is amazing that after Schatz et al. [64] discovered streptomycin in a *S. griseus* strain, about 200 compounds have been reported from other strains identified as *S. griseus*, and still new secondary metabolites are being discovered, pointing to *Actinobacteria* as an inexhaustible source of naturally occurring antibiotics [67].

In a search for bioactive compounds, some authors moved to different or unexplored ecosystems other than soils. In this respect, the exploration of marine organisms is conducted to discover novel bioactive compounds—such as two nucleosides extracted from the sponge *Tectitethya crypta*: spongothymidine and spongouridine [68], or more recently trabectedin, obtained from the tunicate *Ecteinascidia turbinata* [69]. Another approach was to increase the efforts in screening rare *Actinobacteria* genera [70] or explore unknown niches [71].

In the biosphere there is a little-explored niche which can provide promising results: the subterranean environment. The study of the microbiology of caves is an interesting line of research and allows the search of a great diversity of unknown bacteria and fungi and the possible production of new bioactive compounds, as reported by Cheeptham [72] and Cheeptham and Saiz-Jimenez [73].

Caves are colonized by complex bacteria communities and are an excellent reservoir for new species of *Actinobacteria* [71,74,75]. Research on this topic has not been considered with all the dedication it deserves, to the point that the microbiology of most caves is relatively unknown.

Nowadays, two main approaches are adopted in the study of cave bacteria. One is the isolation of the organism and the subsequent assay of the antimicrobial activity of extracts against pathogenic bacteria. Numerous studies can be found in the literature, for which only a few representative papers are cited here [76–81]. In most papers, attempts to identify the bioactive compounds through chemical and structural analyses were not accomplished. The screening of only antimicrobial activity without a structural elucidation of the involved metabolites may not be useful for the discovery of new antibiotics.

A few examples of studies where the bioactive compounds were fully identified are those of Herold et al. [82], who provided the chemical structure of cervimycin A-D, a polyketide glycoside complex obtained from *Streptomyces tendae*, isolated from Grotta dei Cervi, in Italy, and the huanglongmycin A-C complex, synthesized by a strain of *Streptomyces* found in a Chinese cave [83].

Axenov-Gibanov et al. [84] isolated a *Streptomyces* sp. from a cave moonmilk which produced cyclodysidin D and chaxalactin B. Another 120 metabolites were observed in the liquid extract, from which a total of 102 compounds could not be identified and appeared to be novel. These three cases exemplify how caves are a research field with high potential for novel drug discovery.

In addition to conventional isolation techniques, several authors explored the possibilities of increasing the recovery of novel or interesting bacteria by improving isolation and cultivation techniques in order to extend the number of bacteria producing bioactive compounds under laboratory conditions [85]. Some of these methodologies include pretreating samples under different conditions—air drying, dry heating, moist incubation, desiccation, setting different pH, design of new culture media mimicking nature conditions, etc.—for an effective isolation of rare *Actinobacteria* [71,86,87]. A list of non-specific and specific methods that enhance the production of bioactive compounds can be found in Manteca and Yagüe [88].

Another approach is represented by NGS techniques in combination with genome mining which revolutionized the field of antibiotic research. According to Nett et al. [89] genome mining analyses suggest that less than 10% of the genetic potential of antibiotic producers is currently being used, which indicates that there is a huge untapped genetic reservoir waiting to be exploited for drug discovery.

Currently, more than 1555 completed genome sequences of *Streptomyces* are available in EzBiocloud [90] and, for instance, *Streptomyces coelicolor* harbors 22 secondary metabolite gene clusters but really produces only four of the encoded metabolites under standard laboratory conditions [58].

Screening for biosynthetic genes is an effective strategy to characterize bioactivity. For Bukelskis et al. [91] information on the expression of biosynthetic genes encoding for various bioactive compounds in cave bacteria is either limited or missing, and genome mining for PKS and NRPS genes in parallel with transcriptional analysis of the identified genes would be the more effective strategy to analyze and exploit the bioactivity of cave bacterial strains.

Along with genomic mining, the combination of culturing techniques and transcriptomics would complement the systematic investigation for bioactive compounds in bacteria. These techniques have been recently used to investigate the production of the antibiotic andrimid in the marine bacterium *Vibrio coralliilyticus* [92]. In this study, the authors reported the differential expression of five biosynthetic gene clusters as well as an overproduction of andrimid in the presence of chitin rather than glucose in the culture medium. Alteration of cultivation parameters, such as, solid/liquid culturing, the presence/absence of nutrients, variations of pH and temperature, or changes in aeration supplying, would lead to the activation/inactivation of metabolic pathways involved in the biosynthesis of antimicrobials. These conditions or phenotypic variations are used in transcriptomic analyses to identify the clusters of genes differentially expressed in every condition, and afterwards, expressed in heterologous hosts by means of vectors such as plasmids, cosmids, fosmids, and artificial bacterial chromosomes or bioactive compounds. Methodology based on heterologous expression and subsequent screening has been widely implemented in several species of the genus *Streptomyces* and other actinobacteria for secondary metabolism expression and subsequent identification of bioactive compounds [93].

Culturing and isolation of microorganisms has been the primary methodology in new antibiotics discovery but, although nowadays it is extensively used, this approach is biased by the impossibility of extrapolating the subsurface microbiome to the laboratory. In fact, the majority of bacterial species cannot grow in laboratory conditions, an issue that could have led to disregarding a huge amount of new antibiotics synthetized by uncultured bacteria [87]. To amend the culture-dependent bias, metagenomic mining and metatranscriptomics should raise the chance to acquire novel bioactive compounds directly from the subterranean ecosystems.

Metagenomics allows the study of genomes from non-culturable bacteria by means of sequencing the in-situ collected DNA and the subsequent bioinformatic analyses for assembly, binning, and annotation of the genomes of bacteria present in the sample. As a result of the treatment of metagenomic data, potential biosynthetic gene clusters are set in libraries and expressed by heterologous hosts to evaluate and validate their bioactivity.

Metatranscriptomics have a double impact in microbiome studies—on one hand, RNA sequencing can focus on the metabolically active bacteria, discriminating the ancient DNA and latent or inactive bacteria that could add "noise" to the study, since the variation of relative abundance and presence of bacterial communities has been observed, even at the phylum level, in rRNA 16S studies when comparing total and metabolically active communities using both cloning and NGS analyses [33,94]. Thus, efforts of bioactivity analyses could be more exhaustive and accurate. On the other hand, interactions inter- and intraspecies are the best frames to investigate the bioactivity of bacteria in competition with other species sharing the same niche.

In the last decade, a few projects have been funded by the European Commission and National Organizations. They are SeaBioTech (https://spider.science.strath.ac.uk/seabiotech/index.php), Marex (https://www.marex.fi/), PharmaSea (http://www.pharma-sea.eu/pharmasea/), FucoSan (https://www.fucosan.eu/en/), Tascmar (https://cordis.europa.eu/project/id/634674), and Probio (https://www.vliz.be/en/news?p=show&id=8386). All of them focused their activities on marine organisms (bacteria, algae, invertebrates, etc.). Activities and results can be found in their respective web pages. As far as we know no project on terrestrial microorganisms is ongoing, other than a recent research project launched with the aim of studying the biodiversity of extreme environments such as abandoned and active mines in Euroregion A3 (Alentejo, Algarve, and Andalucia) [95].

The mines (pyrite, manganese, copper, etc.) to be investigated are located in the Iberian Pyrite Belt. The Iberian Pyrite Belt is one of the World's largest accumulations of mine wastes and acidic mine waters from drainages, which have caused severe pollution by the low pH and presence of dissolved metals. This acidity and metal pollution has caused the loss of most forms of aquatic life, with the exception of acidophilic microorganisms which inhabit these extreme environments [96]. In this scenario, microorganisms are subjected to stress and the need to develop a metabolic system capable of coping with oligotrophy (lack of organic nutrients) and the expression of genes that produce bioactive compounds to compete with other organisms for the scarce nutrients available in the acidic environment.

The inclusion of mines in this project represents a further step and an innovative research area, since as far as we know there are no previous reports considering mine microbiomes as a source of bioactive compounds.

Preliminary results (Table 1) showed that about 14% of the bacteria isolated from caves and mines produced bioactive compounds against the tested pathogens. It was noteworthy that bacteria isolated from sediments of two submarine cave and marine organisms inhabiting the caves, located in the Algarve coast, only yielded two positive strains, in contrast with the higher number of terrestrial isolates. In addition, it is remarkable that 29% of bacteria isolated from the Iberian Pyrite Belt mines showed inhibitory activity against pathogens, which doubled the percentage of cave bacteria. Most of the bacteria showed high activity against Gram-positive bacteria and lower against Gram-negative bacteria. This is a reason to focus the search on having antibiotics active against Gram-negative bacteria, as demanded by the WHO.

Table 1. Screening of bacteria from mines and marine caves in Alentejo, Algarve, and Andalucia, as well as from the IRNAS-CSIC cave bacterial collection.

Origin	Tested Strains	Positive Strains	1	2	3	4	5	6
Terrestrial caves	863	126	77	48	116	18	31	24
Marine caves	144	2	0	0	2	0	1	0
Mines	79	23	13	7	14	3	8	3
Total (bacteria)	1086	151	90	55	132	21	40	27
Total (%)	100%	13.9%	8.29%	5.06%	12.15%	1.93%	3.68%	2.49%

1: *Bacillus cereus* CECT 148; 2: *Staphylococcus aureus* CECT 4630; 3: *Arthrobacter* sp. LR584284; 4: *Pseudomonas aeruginosa* CECT 110; 5: *Escherichia coli* DSM 105182; 6: *Acinetobacter baumannii* DSM 300007.

A complete screening of bacteria from air, water, and sediments from mines provided a bacterium collected with an air sampler [97] inside the mine of Lousal (Portugal). Based on the study of the 16S rRNA gene and subsequent sequencing of the genome, the strain was identified as a species within the genus *Pseudomonas* and showed a high bioactivity against all the tested pathogenic bacteria. Genome mining analysis focused on the secondary metabolism with antiSMASH [98] which resulted in the prediction of 19 biosynthetic gene clusters from different domains such as polyketide synthases, non-ribosomal peptide synthases, tRNA-dependent cyclodipeptide synthases, aryl polyenes, phenazines, bacteriocins, N-acetylglutaminylglutamine amides, betalactones, flavin-dependent tryptophan halogenases, homoserine lactones, and siderophores (Figure 3).

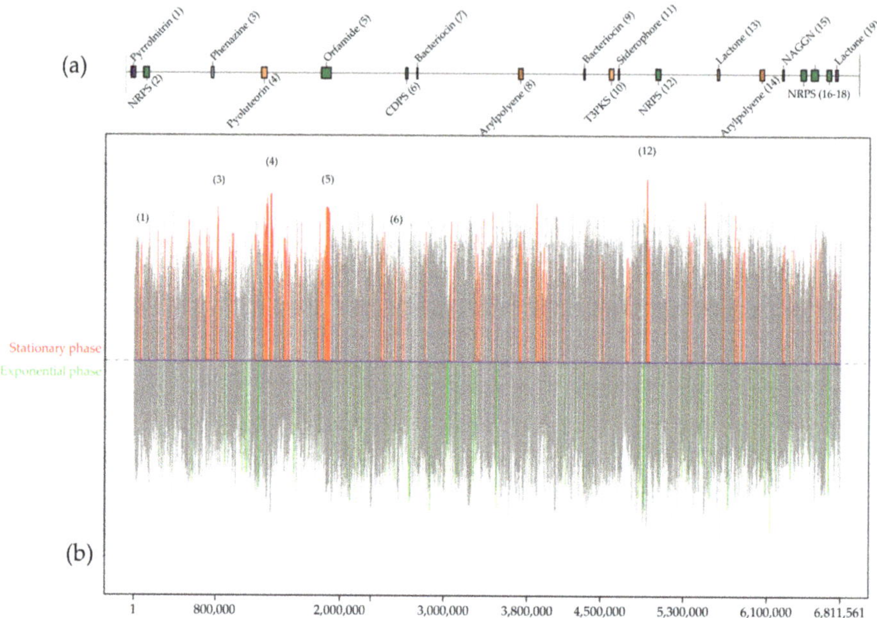

Figure 3. Graphic example from a differential gene expression analysis for *Pseudomonas* sp., isolated from a mine in Lousal (Portugal). Culturing conditions were established in solid medium with nutrient agar and 2% of glycerol. The prediction and location of the biosynthetic gene clusters in the 6.8 Mbp genome (**a**) were used to map the differentially expressed genes (**b**) between the stationary phase and the exponential along the chromosome. Colored peaks represent the variability of expressed genes between both conditions with a 99% of probability. Red peaks show those genes differentially expressed in the stationary phase, in opposition to the green peaks, which are referred to the differential expression in the exponential phase. Grey peaks represent those genes differentially expressed below the 99% threshold, or simply not differentially expressed between conditions. From the predicted secondary metabolites, only six of them were differentially expressed in the stationary phase. Considering the bioactivity analyses in this medium and the study of antibiotic resistance mechanisms on genomes of pathogenic bacteria, no expression of genes involved in the synthesis of antibiotics against Gram-negative bacteria occurred for the described solid media culture.

An "One Strain Many Compounds" (OSMAC) approach [99] was used to check the bioactivity of *Pseudomonas* sp. with the aim of performing a comparative analysis of gene expression by means of transcriptomics techniques. A variation of pattern in the inhibition of pathogens was observed between solid and liquid cultures. The bioactivity of *Pseudomonas* sp. was limited to Gram-positive bacteria in a solid culture with nutrient agar and 2% of glycerol, whereas for the liquid version of the same culture, the inhibition of pathogens was total, for both Gram-positive and Gram-negative bacteria. Inhibition of growth in pathogens was only observed during the stationary phase. Thus, prediction of gene clusters involved in the synthesis of secondary metabolites, comparison of analyses based on the gene expressions, and the study of the antibiotic's resistance mechanisms, or resistome, from the pathogenic bacteria genomes allowed the identification of specific genetic mechanisms involved in the antimicrobial bioactivity for the tested *Pseudomonas* sp. strain.

For a correct management and understanding of genomics and transcriptomics data, advances of bioinformatics have empowered the development of these omics sciences. Software tools addressing quality checking, correction, assembly, mapping, and annotation of data represent the basic procedure, but the availability of hardware with the capacity to carry out high-level operations is essential. However, these analyses and processes are harder and tedious for functional metagenomics and metatranscriptomics because of a higher presence of data coming from hundred or even thousand bacterial genomes present in the same sample. In this sense, the continuous improvement of bioinformatics tools and workflows are needed to achieve a deeper understanding of biological processes [100].

The improvement in NGS platforms and the rising use of omics sciences has stimulated the falling cost of sequencing and the appearance of better computational methodologies. Nevertheless, although these techniques are presented as a prominent stimulus in new antibiotics research, few studies have focused on cave microbiome have used metagenomic and metatranscriptomic analyses so far. Beyond the inherent bias to the application of nucleic acid-based techniques such as a low concentration of DNA/RNA, partial fragmentation, and presence of enzyme inhibitors [101], current limitations are focused on the training of computational scientists to analyze the complex data as well as sequencing enough samples to get a powerful study [102].

4. Conclusions

Subterranean ecosystems constitute a huge subsurface reactor of the global biogeochemical cycle with a potential and regular buffering effect on long-term increments of atmospheric GHGs linked to climate change. The subterranean microbiome plays a primary regulatory role in its gas composition, controlling the uptake-fixation-production of CO_2, CH_4 and N_xO_x gases, as well as their coupled evolution during their migration through the critical zone to the lower troposphere. Recent studies apply an innovative and multidisciplinary combination of a broad suite of cutting-edge technologies -GHG flux monitoring, isotopic geochemical tracing, biogeochemistry, metagenomics, etc.—to quantify the GHG fluxes controlled by microbial-induced processes and directly exchanged with the cave atmosphere in several temporal scales (daily, seasonal, annual pattern). The next step should be to establish the feedback mechanisms between environmental-microclimatic conditions and rates and type of activity of microbial communities in subterranean ecosystems.

The recent in-depth study on the microbial activity of the subterranean microbiota has not only provided qualitative and quantitative data on the regulation of the concentration of CO_2 and CH_4, but has also shown that methanotrophs and heterotrophs can interact and stimulate the growth of each other. This growth results in the production of bioactive compounds. Although the literature describes the isolation of many cave bacteria with inhibitory properties against pathogens, only a few studies provide the identification and chemical structure of the metabolites. The advent of massive sequencing technologies or NGS has promoted the development of the so-called omics sciences, offering a holistic view of the inter- and intracommunity relationships of microorganisms. Genomics, transcriptomics, and the consequent massive generation of data have favored the development of bioinformatics as an essential interdisciplinary field, in continuous progress, for the interpretation of biological processes. This set of techniques and disciplines has not only allowed a more exhaustive knowledge of the biology of microorganisms, but it has made possible to overcome the barrier of extrapolating the ideal conditions for microbial growth in the natural environment to the laboratory, and in this way it has also been made possible to identify new genetic mechanisms involved in the synthesis of bioactive compounds.

Author Contributions: Conceptualization, S.S.-M., S.C. and C.S.-J.; investigation, T.M.-P., S.C., V.J., J.L.G.-P., I.D.-M. and J.C.C.; writing—original draft preparation, T.M.-P., S.S.-M., J.L.G.-P., A.F.-C. and C.S.-J.; writing—review and editing, C.S.-J. All authors have read and agreed to the published version of the manuscript.

Funding: Financial support was obtained through project 0483_PROBIOMA_5_E, co-financed by the European Regional Development Fund within the framework of the Interreg V-A Spain-Portugal program (POCTEP) 2014–2020. This work was also supported by the Spanish Ministry of Economy and Competitiveness through projects CGL2016-75590-P and PID2019-110603RB-I00, AEI/FEDER, UE.

Conflicts of Interest: The authors declare no conflict of interest.

References

1. Ford, D.C.; Williams, P.W. *Karst Hydrogeology and Geomorphology*; Wiley: Chichester, UK, 2007.
2. Giardino, J.R.; Houser, C. (Eds.) Introduction to the critical zone. In *Principles and Dynamics of the Critical Zone*; Elsevier: Amsterdam, The Netherlands, 2015; pp. 1–13.
3. Gold, T. The deep, hot biosphere. *Proc. Natl. Acad. Sci. USA* **1992**, *89*, 6045–6049. [CrossRef] [PubMed]
4. Jones, A.A.; Bennett, P.C. Mineral ecology: Surface specific colonization and geochemical drivers of biofilm accumulation, composition, and phylogeny. *Front. Microbiol.* **2017**, *8*, 491. [CrossRef] [PubMed]
5. Cuezva, S.; Sanchez-Moral, S.; Saiz-Jimenez, C.; Cañaveras, J.C. Microbial communities and associated mineral fabrics in Altamira Cave, Spain. *Int. J. Speleol.* **2009**, *38*, 83–92. [CrossRef]
6. Northup, D.E.; Lavoie, K.H. Geomicrobiology of caves: A review. *Geomicrobiol. J.* **2001**, *18*, 199–222. [CrossRef]
7. Sánchez-Moral, S.; Portillo, M.D.C.; Janices, I.; Cuezva, S.; Cortés, Ángel, F.; Cañaveras, J.C.; Gonzalez, J.M. The role of microorganisms in the formation of calcitic moonmilk deposits and speleothems in Altamira Cave. *Geomorphology* **2012**, *139–140*, 285–292. [CrossRef]
8. Cuezva, S.; Fernandez-Cortes, A.; Porca, E.; Pašić, L.; Jurado, V.; Hernández, M.; Serrano-Ortiz, P.; Hermosin, B.; Cañaveras, J.C.; Sanchez-Moral, S.; et al. The biogeochemical role of Actinobacteria in Altamira Cave, Spain. *FEMS Microbiol. Ecol.* **2012**, *81*, 281–290. [CrossRef] [PubMed]
9. Jones, R.M.; Goordial, J.M.; Orcutt, B.N. Low energy subsurface environments as extraterrestrial analogs. *Front. Microbiol.* **2018**, *9*, 1605. [CrossRef]
10. Lavoie, K.H.; Winter, A.S.; Read, K.J.H.; Hughes, E.M.; Spilde, M.N.; Northup, D.E. Comparison of bacterial communities from lava cave microbial mats to overlying surface soils from Lava Beds National Monument, USA. *PLoS ONE* **2017**, *12*, e0169339. [CrossRef]
11. Miller, A.Z.; Garcia-Sanchez, A.M.; Martin-Sanchez, P.M.; Pereira, M.F.C.; Spangenberg, J.E.; Jurado, V.; Dionísio, A.; Afonso, M.J.; Chaminé, H.I.I.S.; Hermosin, B.; et al. Origin of abundant moonmilk deposits in a subsurface granitic environment. *Sedimentology* **2018**, *65*, 1482–1503. [CrossRef]
12. Sasowsky, I.D.; Mylroie, J. *Studies of Cave Sediments. Physical and Chemical Records of Paleoclimate*; Springer: Dordrecht, The Netherlands, 2012.
13. Ghosh, S.; Kuisiene, N.; Cheeptham, N. The cave microbiome as a source for drug discovery: Reality or pipe dream? *Biochem. Pharmacol.* **2017**, *134*, 18–34. [CrossRef]
14. Engel, A.S. *Microbial Life of Cave Systems*; De Gruiter: Berlin, Germany, 2015.
15. Moldovan, O.T.; Kovac, L.; Halse, S. *Cave Ecology*; Springer: Cham, Switzerland, 2018.
16. Jones, B. Microbial activity in caves—A geological perspective. *Geomicrobiol. J.* **2001**, *18*, 345–357. [CrossRef]
17. Barton, H.A.; Northup, D.E. Geomicrobiology in cave environments: Past, current and future perspectives. *J. Cave Karst Stud.* **2007**, *69*, 163–178.
18. Engel, A.S. Microbial diversity of cave ecosystems. In *Geomicrobiology: Molecular and Environmental Perspectives*; Barton, L.L., Mandl, M., Loy, A., Eds.; Springer: Dordrecht, The Netherlands, 2010; pp. 219–238.
19. Tomczyk-Żak, K.; Zielenkiewicz, U. Microbial diversity in caves. *Geomicrobiol. J.* **2016**, *33*, 20–38. [CrossRef]
20. Jones, D.S.; Macalady, J.L. The snotty and the stringy: Energy for subsurface life in caves. In *Their World: A Diversity of Microbial Environments*; Hurst, C.J., Ed.; Springer: Cham, Switzerland, 2016; pp. 203–224.
21. De Mandal, S.; Chatterjee, R.; Kumar, N.S. Dominant bacterial phyla in caves and their predicted functional roles in C and N cycle. *BMC Microbiol.* **2017**, *17*, 90. [CrossRef]
22. Etiope, G.; Klusman, R.W. Geologic emissions of methane to the atmosphere. *Chemosphere* **2002**, *49*, 777–789. [CrossRef]
23. Etiope, G.; Lollar, B.S. Abiotic methane on earth. *Rev. Geophys.* **2013**, *51*, 276–299. [CrossRef]
24. Intergovernmental Panel on Climate Change. Carbon and other biogeochemical cycles. In *Climate Change 2013: The Physical Science Basis*; Cambridge Univ. Press: Cambridge, UK, 2014; pp. 465–570.

25. Hall, E.K.; Bernhardt, E.S.; Bier, R.L.; Bradford, M.A.; Boot, C.M.; Cotner, J.B.; Del Giorgio, P.A.; Evans, S.E.; Graham, E.B.; Jones, S.E.; et al. Understanding how microbiomes influence the systems they inhabit. *Nat. Microbiol.* **2018**, *3*, 977–982. [CrossRef]
26. Op den Camp, H.J.M.; Islam, T.; Stott, M.B.; Harhangi, H.R.; Hynes, A.; Schouten, S.; Jetten, M.S.M.; Birkeland, N.-K.; Pol, A.; Dunfield, P.F. Environmental, genomic and taxonomic perspectives on methanotrophic *Verrucomicrobia*. *Environ. Microbiol. Rep.* **2009**, *1*, 293–306. [CrossRef]
27. Butterfield, C.N.; Li, Z.; Andeer, P.F.; Spaulding, S.; Thomas, B.C.; Singh, A.; Hettich, R.L.; Suttle, K.B.; Probst, A.J.; Tringe, S.G.; et al. Proteogenomic analyses indicate bacterial methylotrophy and archaeal heterotrophy are prevalent below the grass root zone. *PeerJ* **2016**, *4*, e2687. [CrossRef]
28. Hutchens, E.; Radajewski, S.; Dumont, M.G.; McDonald, I.R.; Murrell, J.C. Analysis of methanotrophic bacteria in Movile Cave by stable isotope probing. *Environ. Microbiol.* **2004**, *6*, 111–120. [CrossRef]
29. Chen, Y.; Wu, L.; Boden, R.; Hillebrand, A.; Kumaresan, D.; Moussard, H.; Baciu, M.; Lu, Y.; Murrell, J.C. Life without light: Microbial diversity and evidence of sulfur- and ammonium-based chemolithotrophy in Movile Cave. *ISME J.* **2009**, *3*, 1093–1104. [CrossRef] [PubMed]
30. Kumaresan, D.; Wischer, D.; Stephenson, J.; Hillebrand-Voiculescu, A.; Murrell, J.C. Microbiology of Movile Cave—A chemolithoautotrophic ecosystem. *Geomicrobiol. J.* **2014**, *31*, 186–193. [CrossRef]
31. Kumaresan, D.; Stephenson, J.; Doxey, A.C.; Bandukwala, H.; Brooks, E.; Hillebrand-Voiculescu, A.; Whiteley, A.S.; Murrell, J.C. Aerobic proteobacterial methylotrophs in Movile Cave: Genomic and metagenomic analyses. *Microbiome* **2018**, *6*, 1–10. [CrossRef] [PubMed]
32. Saiz-Jimenez, C. The microbiology of show caves, mines tunnels and tombs: Implications for management and conservation. In *Microbial Life of Cave Systems*; Engel, A.S., Ed.; De Gruiter: Berlin, Germany, 2015; pp. 231–261.
33. Gonzalez-Pimentel, J.L.; Miller, A.Z.; Jurado, V.; Laiz, L.; Pereira, M.F.C.; Saiz-Jimenez, C. Yellow coloured mats from lava tubes of La Palma (Canary Islands, Spain) are dominated by metabolically active Actinobacteria. *Sci. Rep.* **2018**, *8*, 1944. [CrossRef] [PubMed]
34. Fernandez-Cortes, A.; Cuezva, S.; Alvarez-Gallego, M.; Garcia-Anton, E.; Pla, C.; Benavente, D.; Jurado, V.; Saiz-Jimenez, C.; Sanchez-Moral, S. Subterranean atmospheres may act as daily methane sinks. *Nat. Commun.* **2015**, *6*, 7003. [CrossRef] [PubMed]
35. McDonald, I.R.; Murrell, J.C. The particulate methane monooxygenase gene pmoA and its use as a functional gene probe for methanotrophs. *FEMS Microbiol. Lett.* **1997**, *156*, 205–210. [CrossRef]
36. Tveit, A.T.; Hestnes, A.G.; Robinson, S.L.; Schintlmeister, A.; Dedysh, S.N.; Jehmlich, N.; Von Bergen, M.; Herbold, C.; Wagner, M.; Richter, A.; et al. Widespread soil bacterium that oxidizes atmospheric methane. *Proc. Natl. Acad. Sci. USA* **2019**, *116*, 8515–8524. [CrossRef]
37. Schimmelmann, A.; Fernandez-Cortes, A.; Cuezva, S.; Streil, T.; Lennon, J.T. Radiolysis via radioactivity is not responsible for rapid methane oxidation in subterranean air. *PLoS ONE* **2018**, *13*, e0206506. [CrossRef]
38. Webster, K.D.; Drobniak, A.; Etiope, G.; Mastalerz, M.; Sauer, P.E.; Schimmelmann, A. Subterranean karst environments as a global sink for atmospheric methane. *Earth Planet. Sci. Lett.* **2018**, *485*, 9–18. [CrossRef]
39. Lennon, J.T.; Nguyễn-Thùy, D.; Phạm, T.M.; Drobniak, A.; Tạ, P.H.; Phạm, N.Đ.; Streil, T.; Webster, K.D.; Schimmelmann, A. Microbial contributions to subterranean methane sinks. *Geobiology* **2016**, *15*, 254–258. [CrossRef]
40. Waring, C.L.; Hankin, S.I.; Griffith, D.W.T.; Kertesz, M.A.; Kobylski, V.; Wilson, N.L.; Coleman, N.V.; Kettlewell, G.; Zlot, R.; Bosse, M.; et al. Seasonal total methane depletion in limestone caves. *Sci. Rep.* **2017**, *7*, 1–12. [CrossRef] [PubMed]
41. Ojeda, L.; Vadillo, I.; Etiope, G.; Benavente, J.; Liñán, C.; del Rosal, Y.; Tapia, S.T.; Moríñigo, M.Á.; Carrasco, F. Methane sources and sinks in karst systems: The Nerja cave and its vadose environment (Spain). *Geochim. Cosmochim. Acta* **2019**, *259*, 302–315. [CrossRef]
42. Cuezva, S.; Martin-Pozas, T.; Fernandez-Cortes, A.; Cañaveras, J.C.; Janssens, I.; Sanchez-Moral, S. On the role of cave-soil in the carbon cycle. A first approach. In Proceedings of the EGU General Assembly 2020, Online, 4–8 May 2020. Abstract 21793. [CrossRef]
43. Jurado, V.; Gonzalez-Pimentel, J.L.; Miller, A.Z.; Hermosin, B.; D'Angeli, I.M.; Tognini, P.; De Waele, J.; Saiz-Jimenez, C. Microbial communities in vermiculation deposits from an Alpine cave. *Front Earth Sci.* **2020**, in press.

44. Martin-Pozas, T.; Sánchez-Moral, S.; Cuezva, S.; Jurado, V.; Saiz-Jimenez, C.; López, R.P.; Carrey, R.; Otero, N.; Giesemann, A.; Well, R.; et al. Biologically mediated release of endogenous N2O and NO2 gases in a hydrothermal, hypoxic subterranean environment. *Sci. Total Environ.* **2020**, *747*, 141218. [CrossRef] [PubMed]
45. Fernandez-Cortes, A.; Perez-Lopez, R.; Cuezva, S.; Calaforra, J.M.; Cañaveras, J.C.; Sanchez-Moral, S. Geochemical fingerprinting of rising deep endogenous gases in an active hypogenic karst system. *Geofluids* **2018**, *2018*, 1–19. [CrossRef]
46. He, Z.; Cai, C.; Wang, J.; Xu, X.; Zheng, P.; Jetten, M.S.M.; Hu, B. A novel denitrifying methanotroph of the NC10 phylum and its microcolony. *Sci. Rep.* **2016**, *6*, srep32241. [CrossRef]
47. Isobe, K.; Bouskill, N.J.; Brodie, E.L.; Sudderth, E.A.; Martiny, J.B.H. Phylogenetic conservation of soil bacterial responses to simulated global changes. *Philos. Trans. R. Soc. B* **2020**, *375*, 20190242. [CrossRef]
48. Cappelletti, M.; Ghezzi, D.; Zannoni, D.; Capaccioni, B.; Fedi, S. Diversity of methane-oxidizing bacteria in soils from "Hot Lands of Medolla" (Italy) featured by anomalous high-temperatures and biogenic CO_2 emission. *Microbes Environ.* **2016**, *31*, 369–377. [CrossRef]
49. Kenney, G.E.; Rosenzweig, A.C. Methanobactins: Maintaining copper homeostasis in methanotrophs and beyond. *J. Biol. Chem.* **2018**, *293*, 4606–4615. [CrossRef]
50. Jiang, H.; Chen, Y.; Jiang, P.; Zhang, C.; Smith, T.J.; Murrell, J.C.; Xing, X.-H. Methanotrophs: Multifunctional bacteria with promising applications in environmental bioengineering. *Biochem. Eng. J.* **2010**, *49*, 277–288. [CrossRef]
51. Pandey, V.C.; Singh, J.S.; Singh, D.P.; Singh, R.P. Methanotrophs: Promising bacteria for environmental remediation. *Int. J. Environ. Sci. Technol.* **2014**, *11*, 241–250. [CrossRef]
52. Iguchi, H.; Yurimoto, H.; Sakai, Y. Stimulation of methanotrophic growth in cocultures by cobalamin excreted by Rhizobia. *Appl. Environ. Microbiol.* **2011**, *77*, 8509–8515. [CrossRef] [PubMed]
53. Stock, M.; Hoefman, S.; Kerckhof, F.-M.; Boon, N.; De Vos, P.; De Baets, B.; Heylen, K.; Waegeman, W. Exploration and prediction of interactions between methanotrophs and heterotrophs. *Res. Microbiol.* **2013**, *164*, 1045–1054. [CrossRef] [PubMed]
54. Veraart, A.J.; Garbeva, P.; Van Beersum, F.; Ho, A.; Hordijk, C.A.; Meima-Franke, M.; Zweers, A.J.; Bodelier, P.L.E. Living apart together—Bacterial volatiles influence methanotrophic growth and activity. *ISME J.* **2018**, *12*, 1163–1166. [CrossRef]
55. World Health Organization. *Global Action Plan on Antimicrobial Resistance*; World Health Organization: Geneva, Switzerland, 2015.
56. De Kraker, M.E.; Stewardson, A.J.; Harbarth, S. Will 10 million people die a year due to antimicrobial resistance by 2050? *PLoS Med.* **2016**, *13*, e1002184. [CrossRef]
57. World Health Organization. *Prioritization of Pathogens to Guide Discovery, Research and Development of New Antibiotics for Drug-Resistant Bacterial Infections, Including Tuberculosis*; World Health Organization: Geneva, Switzerland, 2017.
58. Tacconelli, E.; Carrara, E.; Savoldi, A.; Harbarth, S.; Mendelson, M.; Monnet, D.L.; Pulcini, C.; Kahlmeter, G.; Kluytmans, J.; Carmeli, Y.; et al. Discovery, research, and development of new antibiotics: The WHO priority list of antibiotic-resistant bacteria and tuberculosis. *Lancet Infect. Dis.* **2018**, *18*, 318–327. [CrossRef]
59. World Health Organization. *Antibacterial Agents in Clinical Development: An Analysis of the Antibacterial Clinical Development Pipeline*; World Health Organization: Geneva, Switzerland, 2019.
60. Theuretzbacher, U. Antibiotic innovation for future public health needs. *Clin. Microbiol. Infect.* **2017**, *23*, 713–717. [CrossRef]
61. Barka, E.A.; Vatsa, P.; Sanchez, L.; Gaveau-Vaillant, N.; Jacquard, C.; Klenk, H.-P.; Clément, C.; Ouhdouch, Y.; Van Wezel, G.P. Taxonomy, physiology, and natural products of *Actinobacteria*. *Microbiol. Mol. Biol. Rev.* **2015**, *80*, 1–43. [CrossRef]
62. Mast, Y.; Stegmann, E. Actinomycetes: The antibiotics producers. *Antibiotics* **2018**, *8*, 105. [CrossRef]
63. Waksman, S.A.; Woodruff, H.B. Bacteriostatic and bactericidal substances produced by a soil Actinomyces. *Exp. Biol. Med.* **1940**, *45*, 609–614. [CrossRef]
64. Schatz, A.; Bugle, E.; Waksman, S.A. Streptomycin, a substance exhibiting antibiotic activity against gram-positive and gram-negative bacteria. *Exp. Biol. Med.* **1944**, *55*, 66–69. [CrossRef]
65. Nakashima, T.; Anzai, K.; Suzuki, R.; Kuwahara, N.; Takeshita, S.; Kanamoto, A.; Ando, K. Productivity of bioactive compounds in *Streptomyces* species isolated from Nagasaki marine environments. *Actinomycetologica* **2009**, *23*, 16–20. [CrossRef]

66. Koomsiri, W.; Inahashi, Y.; Kimura, T.; Shiomi, K.; Takahashi, Y.; Ōmura, S.; Thamchaipenet, A.; Nakashima, T. Bisoxazolomycin A: A new natural product from '*Streptomyces subflavus* subsp. irumaensis' AM-3603. *J. Antibiot.* **2017**, *70*, 1142–1145. [CrossRef]
67. Takahashi, Y.; Nakashima, T. Actinomycetes, an inexhaustible source of naturally occurring antibiotics. *Antibiotics* **2018**, *7*, 45. [CrossRef] [PubMed]
68. Bergmann, W.; Burke, D.C. Marine products. XXXIX. The nucleosides of sponges. III. Spongothymidine and spongouridine. *J. Org. Chem.* **1955**, *20*, 1501–1507. [CrossRef]
69. Lindequist, U. Marine-derived pharmaceuticals—Challenges and opportunities. *Biomol. Ther.* **2016**, *24*, 561–571. [CrossRef]
70. Tiwari, K.; Gupta, R.K. Diversity and isolation of rare actinomycetes: An overview. *Crit. Rev. Microbiol.* **2013**, *39*, 256–294. [CrossRef]
71. Fang, B.-Z.; Salam, N.; Han, M.-X.; Jiao, J.-Y.; Cheng, J.; Wei, D.-Q.; Xiao, M.; Li, W.-J. Insights on the effects of heat pretreatment, pH, and calcium salts on isolation of rare *Actinobacteria* from karstic caves. *Front. Microbiol.* **2017**, *8*, 1535. [CrossRef]
72. Cheeptham, N. *Cave Microbiomes: A Novel Resource for Drug Discovery*; Springer: New York, NY, USA, 2013.
73. Cheeptham, N.; Saiz-Jimenez, C. New sources of antibiotics: Caves. In *Antibiotics. Current Innovations and Future Trends*; Sánchez, S., Demain, A.L., Eds.; Caister Academic Press: Portland, Oregon, 2015; pp. 213–227.
74. Groth, I.; Vettermann, R.; Schuetze, B.; Schumann, P.; Saiz-Jimenez, C. Actinomycetes in Karstic caves of northern Spain (Altamira and Tito Bustillo). *J. Microbiol. Methods* **1999**, *36*, 115–122. [CrossRef]
75. Jurado, V.; Laiz, L.; Rodríguez-Nava, V.; Boiron, P.; Hermosin, B.; Sanchez-Moral, S.; Saiz-Jimenez, C. Pathogenic and opportunistic microorganisms in caves. *Int. J. Speleol.* **2010**, *39*, 15–24. [CrossRef]
76. Cheeptham, N.; Sadoway, T.; Rule, D.; Watson, K.; Moote, P.; Soliman, L.C.; Azad, N.; Donkor, K.K.; Horne, D. Cure from the cave: Volcanic cave actinomycetes and their potential in drug discovery. *Int. J. Speleol.* **2013**, *42*, 35–47. [CrossRef]
77. Belyagoubi, L.; Belyagoubi-Benhammou, N.; Jurado, V.; Dupont, J.; Lacoste, S.; Djebbah, F.; Ounadjela, F.Z.; Benaissa, S.; Habi, S.; Abdelouahid, D.E.; et al. Antimicrobial activities of culturable microorganisms (actinomycetes and fungi) isolated from Chaabe Cave, Algeria. *Int. J. Speleol.* **2018**, *47*, 189–199. [CrossRef]
78. Yasir, M. Analysis of bacterial communities and characterization of antimicrobial strains from cave microbiota. *Braz. J. Microbiol.* **2018**, *49*, 248–257. [CrossRef] [PubMed]
79. Rangseekaew, P.; Pathom-Aree, W. Cave Actinobacteria as producers of bioactive metabolites. *Front. Microbiol.* **2019**, *10*, 387. [CrossRef] [PubMed]
80. Avguštin, J.A.; Petrič, P.; Pašić, L. Screening the cultivable cave microbial mats for the production of antimicrobial compounds and antibiotic resistance. *Int. J. Speleol.* **2019**, *48*, 295–303. [CrossRef]
81. Voytsekhovskaya, I.V.; Axenov-Gribanov, D.V.; Murzina, S.A.; Pekkoeva, S.N.; Protasov, E.S.; Gamaiunov, S.V.; Timofeyev, M. Estimation of antimicrobial activities and fatty acid composition of actinobacteria isolated from water surface of underground lakes from Badzheyskaya and Okhotnichya caves in Siberia. *PeerJ* **2018**, *6*, e5832. [CrossRef]
82. Herold, K.; Gollmick, F.A.; Groth, I.; Roth, M.; Menzel, K.-D.; Möllmann, U.; Gräfe, U.; Hertweck, C. Cervimycin A-D: A polyketide glycoside complex from a cave bacterium can defeat vancomycin resistance. *Chem. Eur. J.* **2005**, *11*, 5523–5530. [CrossRef]
83. Jiang, L.; Pu, H.; Xiang, J.; Su, M.; Yan, X.; Yang, D.; Zhu, X.; Shen, B.; Duan, Y.; Huang, Y. Huanglongmycin A-C, cytotoxic polyketides biosynthesized by a putative type ii polyketide synthase from *Streptomyces* sp. CB09001. *Front. Chem.* **2018**, *6*, 254. [CrossRef]
84. Axenov-Gibanov, D.; Voytsekhovskaya, I.V.; Tokovenko, B.T.; Protasov, E.S.; Gamaiunov, S.V.; Rabets, Y.V.; Luzhetskyy, A.N.; Timofeyev, M.A. Actinobacteria isolated from an underground lake moonmilk speleothem from the biggest conglomeratic karstic cave in Siberia as sources of novel biologically active compounds. *PLoS ONE* **2016**, *11*, e0149216. [CrossRef]
85. Bollmann, A.; Lewis, K.; Epstein, S.S. Incubation of environmental samples in a diffusion chamber increases the diversity of recovered isolates. *Appl. Environ. Microbiol.* **2007**, *73*, 6386–6390. [CrossRef]
86. Hug, J.J.; Bader, C.D.; Remškar, M.; Cirnski, K.; Müller, R. Concepts and methods to access novel antibiotics from Actinomycetes. *Antibiotics* **2018**, *7*, 44. [CrossRef] [PubMed]
87. Lewis, K.; Epstein, S.S.; D'Onofrio, A.; Ling, L.L. Uncultured microorganisms as a source of secondary metabolites. *J. Antibiot.* **2010**, *63*, 468–476. [CrossRef] [PubMed]

88. Manteca, A.; Yagüe, P. *Streptomyces* differentiation in liquid cultures as a trigger of secondary metabolism. *Antibiotics* **2018**, *7*, 41. [CrossRef] [PubMed]
89. Nett, M.; Ikeda, H.; Moore, B.S. Genomic basis for natural product biosynthetic diversity in the actinomycetes. *Nat. Prod. Rep.* **2009**, *26*, 1362–1384. [CrossRef]
90. Yoon, S.-H.; Ha, S.-M.; Kwon, S.; Lim, J.; Kim, Y.; Seo, H.; Chun, J. Introducing EzBioCloud: A taxonomically united database of 16S rRNA and whole genome assemblies. *Int. J. Syst. Evol. Microbiol.* **2017**, *67*, 1613–1617. [CrossRef]
91. Bukelskis, D.; Dabkeviciene, D.; Lukoseviciute, L.; Bucelis, A.; Kriaučiūnas, I.; Lebedeva, J.; Kuisiene, N. Screening and transcriptional analysis of polyketide synthases and non-ribosomal peptide synthetases in bacterial strains from Krubera-Voronja Cave. *Front. Microbiol.* **2019**, *10*, 2149. [CrossRef]
92. Giubergia, S.; Phippen, C.; Nielsen, K.F.; Gram, L. Growth on chitin impacts the transcriptome and metabolite profiles of antibiotic-producing *Vibrio coralliilyticus* S2052 and *Photobacterium galatheae* S2753. *mSystems* **2017**, *2*, e00141-16. [CrossRef]
93. Ahmed, Y.; Rebets, Y.; Estévez, M.R.; Zapp, J.; Myronovskyi, M.; Luzhetskyy, A. Engineering of *Streptomyces lividans* for heterologous expression of secondary metabolite gene clusters. *Microb. Cell Fact.* **2020**, *19*, 5. [CrossRef]
94. Paun, V.I.; Icaza, G.; Lavin, P.; Marin, C.; Tudorache, A.; Perşoiu, A.; Dorador, C.; Purcarea, C. Total and potentially active bacterial communities entrapped in a late glacial through holocene ice core from Scarisoara Ice Cave, Romania. *Front. Microbiol.* **2019**, *10*, 1193. [CrossRef]
95. Interreg project. 0483_PROBIOMA_5_E: Prospecting Underground Environments for Microbial Bioactive Compounds with Potential Use in Medicine, Agriculture and Environment. Available online: https://probioma.org (accessed on 15 November 2020).
96. Sánchez España, J. Acid mine drainage in the Iberian Pyrite Belt: An overview with special emphasis on generation mechanisms, aqueous composition and associated mineral phases. *Macla* **2008**, *10*, 34–43.
97. Porca, E.; Jurado, V.; Martin-Sanchez, P.M.; Hermosin, B.; Bastian, F.; Alabouvette, C.; Saiz-Jimenez, C. Aerobiology: An ecological indicator for early detection and control of fungal outbreaks in caves. *Ecol. Indic.* **2011**, *11*, 1594–1598. [CrossRef]
98. Blin, K.; Shaw, S.; Steinke, K.; Villebro, R.; Ziemert, N.; Lee, S.Y.; Medema, M.H.; Weber, T. antiSMASH 5.0: Updates to the secondary metabolite genome mining pipeline. *Nucleic Acids Res.* **2019**, *47*, W81–W87. [CrossRef] [PubMed]
99. Romano, S.; Jackson, S.A.; Patry, S.; Dobson, A.D.W. Extending the "One Strain Many Compounds" (OSMAC) principle to marine microorganisms. *Mar. Drugs* **2018**, *16*, 244. [CrossRef] [PubMed]
100. Jünemann, S.; Kleinbölting, N.; Jaenicke, S.; Henke, C.; Hassa, J.; Nelkner, J.; Stolze, Y.; Albaum, S.P.; Schlüter, A.; Goesmann, A.; et al. Bioinformatics for NGS-based metagenomics and the application to biogas research. *J. Biotechnol.* **2017**, *261*, 10–23. [CrossRef]
101. Alain, K.; Callac, N.; Ciobanu, M.C.; Reynaud, Y.; Duthoit, F.; Jebbar, M. DNA extractions from deep subsea floor sediments: Novel cryogenic mill-based procedure and comparison to existing protocols. *J. Microbiol. Methods* **2011**, *87*, 355–362. [CrossRef]
102. Quince, C.; Walker, A.W.; Simpson, J.T.; Loman, N.J.; Segata, N. Shotgun metagenomics, from sampling to analysis. *Nat. Biotechnol.* **2017**, *35*, 833–844. [CrossRef]

Publisher's Note: MDPI stays neutral with regard to jurisdictional claims in published maps and institutional affiliations.

Article

Occurrence of Fluoroquinolones and Sulfonamides Resistance Genes in Wastewater and Sludge at Different Stages of Wastewater Treatment: A Preliminary Case Study

Damian Rolbiecki [1], Monika Harnisz [1,*], Ewa Korzeniewska [1], Łukasz Jałowiecki [2] and Grażyna Płaza [2]

1 Department of Engineering of Water Protection and Environmental Microbiology, Faculty of Geoengineering, University of Warmia and Mazury in Olsztyn, Prawocheńskiego St. 1, 10-719 Olsztyn, Poland; damian.rolbiecki@uwm.edu.pl (D.R.); ewa.korzeniewska@uwm.edu.pl (E.K.)

2 Department of Environmental Microbiology, Institute for Ecology of Industrial Areas, Kossutha St. 6, 40-844 Katowice, Poland; l.jalowiecki@ietu.pl (Ł.J.); g.plaza@ietu.pl (G.P.)

* Correspondence: monika.harnisz@uwm.edu.pl or monikah@uwm.edu.pl

Received: 9 July 2020; Accepted: 20 August 2020; Published: 22 August 2020

Abstract: This study identified differences in the prevalence of antibiotic resistance genes (ARGs) between wastewater treatment plants (WWTPs) processing different proportions of hospital and municipal wastewater as well as various types of industrial wastewater. The influence of treated effluents discharged from WWTPs on the receiving water bodies (rivers) was examined. Genomic DNA was isolated from environmental samples (river water, wastewater and sewage sludge). The presence of genes encoding resistance to sulfonamides (*sul*1, *sul*2) and fluoroquinolones (*qep*A, *aac(6′)-Ib-cr*) was determined by standard polymerase chain reaction (PCR). The effect of the sampling season (summer – June, fall – November) was analyzed. Treated wastewater and sewage sludge were significant reservoirs of antibiotic resistance and contained all of the examined ARGs. All wastewater samples contained *sul*1 and *aac(6′)-Ib-cr* genes, while the *qep*A and *sul*2 genes occurred less frequently. These observations suggest that the prevalence of ARGs is determined by the type of processed wastewater. The Warmia and Mazury WWTP was characterized by higher levels of the *sul*2 gene, which could be attributed to the fact that this WWTP processes agricultural sewage containing animal waste. However, hospital wastewater appears to be the main source of the *sul*1 gene. The results of this study indicate that WWTPs are significant sources of ARGs, contributing to the spread of antibiotic resistance in rivers receiving processed wastewater.

Keywords: antibiotic resistance; wastewater; WWTP; ARGs; sulfonamides; fluoroquinolones

1. Introduction

The overuse and misuse of antibiotics in human and veterinary medicine, animal farming and agriculture contributes to antibiotic pollution and the spread of antibiotic resistance in the environment [1,2]. The wide use of antimicrobial drugs creates selection pressure, which speeds up microbial evolution and promotes the development of antibiotic resistance mechanisms. The presence of close associations between antimicrobial medicines and resistance has been widely documented around the world [3]. Antibiotics, antibiotic-resistant bacteria (ARB) and antibiotic resistance genes (ARGs) are ubiquitous in surface water [4–6], ground water [7–9] and soil [10–12]. The ongoing spread of these micropollutants in the natural environment contributes to the development of antibiotic resistance mechanisms in environmental bacteria. Antibiotic resistance genes play a fundamental role in this process.

The sources and transmission mechanisms of ARGs have to be investigated in detail to prevent the spread of antibiotic resistance in the environment. Raw municipal and hospital wastewater, treated effluents and sewage sludge are significant reservoirs of ARGs, and they play a significant role in the transfer of antibiotic resistance [1,13,14]. For this reason, the release of ARGs from wastewater treatment plants (WWTPs) and their fate in the environment have attracted considerable research interest in recent years [15–19].

The removal of solids, organics and nutrients is the key process in wastewater treatment. Wastewater treatment plants are not equipped to remove microbiological contaminants, including ARB and ARGs. The activated sludge process is a biological method that is frequently used in wastewater treatment. The conditions inside activated sludge tanks, including high oxygen concentration, high temperature and high densities of bacterial cells, can promote horizontal gene transfer (HGT) between closely related as well as unrelated microorganisms [20]. As a result, treated effluents can contain more copies of a given resistance gene per unit volume than raw wastewater [21,22].

Treated wastewater is evacuated to water bodies, which leads to the release of large gene pools, including ARGs, into the natural environment [18,23]. Receptacles of treated effluents accumulate mobile genetic elements (MGEs) such as plasmids, transposons, insertion sequences and integrons [15,24], which are effective carriers of ARGs. Mobile genetic elements play a key role in the spread of genetic information between even the most phylogenetically distant species of bacteria. Genetic resistance can be further transferred between bacteria, including pathogenic microorganisms, posing a significant threat to human health and life.

Organic solid waste such as sewage sludge is also a considerable source of ARGs, which contribute to the spread of antibiotic resistance in the environment. Sewage sludge treatment leads to the recovery of municipal biosolids. Biosolids are abundant in nutrients, and they are often used as agricultural fertilizers [25]. In Poland, the storage of sewage sludge with a high caloric content was banned in 2016 [26], which increased the supply of sewage sludge for fertilization purposes. The management of sewage sludge delivers unquestioned benefits, but the application of treated sludge in agriculture may also have negative consequences. Research has shown that ARGs levels in treated sewage are very high [27,28] and often significantly exceed ARGs concentrations in wastewater [29,30]. Sewage sludge is usually stabilized before it is used as fertilizer, but according to many authors, biosolids processed with the use of various stabilization methods are still highly abundant in ARGs [30,31]. Elevated concentrations of ARGs in soil samples [28,30,32] and accelerated transmission of ARGs from soil to plants [32] were also reported in farmland fertilized with sewage sludge.

Microorganisms have developed various mechanisms of resistance to antibiotics, depending on the type of antimicrobial drugs and their effects on bacterial cells. In Europe, clinical bacterial isolates are becoming increasingly resistant to fluoroquinolones, which are an important class of antibiotics [33]. Resistance to fluoroquinolones is encoded by *aac(6′)-Ib-cr* and *qep*A genes [34], which are usually localized in plasmid DNA. The *qep*A gene encodes efflux pumps of the major facilitator superfamily (MFS), which are responsible for the transport of antibiotics to extracellular space. The *aac(6′)-Ib-cr* gene encodes an enzyme that suppresses the activity of two fluoroquinolone antimicrobials (ciprofloxacin and norfloxacin) through their acetylation [35].

Sulfonamides are one of the oldest groups of antibiotics used in medicine. In recent years, the use of sulfonamides in human medicine has declined due to growing levels of bacterial resistance [36], but these antimicrobials are still widely applied in veterinary medicine [37]. Sulfonamides inhibit folate synthesis by suppressing the production of dihydropteroate synthase (DHPS) (EC 2.5.1.15), an enzyme involved in the synthesis of folic acid. Bacteria have developed mechanisms of resistance against sulfonamides through mutation of the chromosomal DHPS gene (*folP*) or the acquisition of an alternative DHPS gene (*sul*). Dihydropteroate synthase encoded by a *sul* gene has a low affinity for sulfonamides, and is not inhibited by this class of antibiotics. Three genes encoding resistance to sulfonamides (*sul*1, *sul*2, *sul*3) with estimated 50% sequence similarity have been identified to date [37,38]. It is believed that general resistance to sulfonamides is encoded mostly by *sul*1 and *sul*2.

The acquisition of sulfonamide resistance through *sul* genes is the most prevalent mechanism in the environment [19,39].

This study aimed to determine the prevalence of genes encoding resistance to sulfonamides (*sul*1, *sul*2) and fluoroquinolones (*qep*A, *aac(6')-Ib-cr*) in various stages of wastewater treatment in two WWTPs. The examined plants are situated in the regions of Warmia and Mazury (northern Poland) and Silesia (southern Poland), which differ considerably in industrial development level. The treatment plants use similar methods of wastewater treatment based on activated sludge. Differences in the prevalence of ARGs between the studied WWTPs processing different proportions of the hospital and municipal wastewater as well as various types of industrial wastewater were investigated. The study analyzed whether the inflow of various sources of industrial wastewater (brewery wastewater, animal industry wastewater) affects the differentiation in the occurrence of ARGs in WWTPs. Samples were collected at the subsequent stages of wastewater treatment to capture the ARGs reduction, and the influence of treated effluents evacuated from WWTPs on the receiving water bodies was examined by analyzing samples of river water collected upstream and downstream from wastewater discharge points. Sewage sludge was also analyzed. The impact of the season on the ARGs prevalence was evaluated by analyzing samples collected in summer (June) and fall (November).

2. Materials and Methods

2.1. Study Area and Sampling Sites

In this study, the presence of genes encoding resistance to fluoroquinolones and sulfonamides was analyzed in two WWTPs with similar treatment systems and processing capacities. Two wastewater treatment plants (WWTPs) in Poland were analyzed in the study. The WWTPs are located in different Polish regions: Warmia and Mazury (WM-WWTP) and Silesia (S-WWTP). Treated effluents are evacuated to the Łyna River and the Gostynia River, respectively. Samples of river water collected upstream and downstream from effluent discharge points were examined to determine the influence of wastewater processing technology on rivers receiving treated wastewater. The analyzed WWTPs are characterized by similar treatment technologies, but they differ in the type of inflowing wastewater. Both WWTPs have a similar daily average processing capacity and employ a mechanical-biological treatment system based on the activated sludge. S-WWTP has an average influent flow rate of 32,000 m^3/day, and it receives municipal sewage from the city of Tychy, hospital wastewater and industrial sewage. Industrial wastewater (25% of the receiving wastewater) is supplied mainly by a brewery where sewage is pre-cleaned in the methane fermentation process. The brewery is the largest and the most important industrial facility in the S-WWMT area. S-WWTP deploys mechanical-biological treatment methods and operates sequencing C-TECH reactors, activated sludge chambers (a system of chambers of various oxygen conditions: anaerobic, anoxic and aerobic), a secondary settling tank and an anoxic chamber. S-WWTP uses supplementary chemical phosphorus removal [40]. WM-WW TP operates a mechanical-biological treatment system with an elevated removal of nutrients (MB-ERN) with the following sections of the wastewater treatment process: a pre-denitrification chamber, a phosphorus removal tank, nitrification/denitrification chambers and secondary settling tanks. It processes municipal sewage from the city of Olsztyn, industrial wastewater and sewage from three hospitals. Industrial wastewater (20% of the receiving wastewater) is supplied by animal industry wastewater. According to the information obtained from the administration of the WM-WWTP, the plant has an average processing capacity of 35,000 m^3/day. The prevalence of bacterial infections, the amount of antimicrobial drugs and the number of hospitalized patients differ across seasons. These parameters can influence the results noted in each sampling season. Therefore, samples were collected in the summer (June) and fall (November) of 2018. Samples of river water, wastewater and sludge from various stages of treatment were analyzed (n = 36). The types of samples collected in each WWTP are presented in Table 1.

Table 1. Types of collected samples.

	Region of Silesia (S-WWTP)* (n = 18)		Region of Warmia and Mazury (WM-WWTP)** (n = 18)	
	Type of Sample	**Symbol**	**Type of Sample**	**Symbol**
Liquid samples from WWTPs	Untreated wastewater	S1	Untreated wastewater	W1
	Wastewater from the outlet of the primary clarifier	S2	Wastewater from the outlet of the primary clarifier	W2
	Wastewater from the outlet of the secondary clarifier	S3	Wastewater in the biological chamber	W3
	Wastewater from the outlet of the C-TECH reactor	S4	Wastewater from the outlet of the multipurpose reactor	W4
	Treated wastewater	S5	Treated wastewater	W5
Solid samples from WWTPs	Sludge from the outlet of the mechanical concentrator	S6	Sludge from the outlet of the open fermentation pool	W6
	Sludge from the outlet of the gravity concentrator	S7	Treated sludge	W7
River water	River water upstream the effluent discharge point	S8	River water upstream the effluent discharge point	W8
	River water downstream the effluent discharge point	S9	River water downstream the effluent discharge point	W9

* Silesian wastewater treatment plant; ** Warmia and Mazury wastewater treatment plant.

Three grab samples of around 160 mL of wastewater or river water were individually collected and combined into composite samples in sterile 500 mL bottles. Sludge samples were collected into sterile urine containers. The samples were transported to the laboratory on the day of collection and stored in 4 °C for further analysis.

2.2. DNA Extraction

Samples of wastewater and river water were filtered with the use of vacuum pumps and passed through 0.2 μm pore size polycarbonate membrane filters. The volume of the filtered samples ranged from 10 mL to 400 mL, depending on the site and season of collection. Sludge samples of 0.25 g each were used directly for DNA isolation.

The DNeasy Power Water Kit (Qiagen, Hilden, Germany) was used to isolate genomic DNA from sewage and water samples, and the DNeasy Power Soil Kit (Qiagen, Hilden, Germany) was used to isolate genomic DNA from sewage sludge samples according to the manufacturer's protocol. The quality and quantity of the obtained genetic material was checked with the Multiskan Sky Microplate Spectrophotometer (Thermo Scientific, Waltham, MA, USA).

2.3. Identification of Aantibiotic Resistance Genes by Polymerase Chain Reaction (PCR)

The presence of genes encoding resistance to sulfonamide (*sul1*, *sul2*) and quinolone (*qepA*, *aac (6′)-Ib-cr*) was confirmed by PCR. The applied primers and reaction profiles are presented in Table 2. The PCR reactions were performed in a volume of 15 μL containing 10 μM of the respective primer pairs, 1 μL of genomic DNA of each sample and the NZYTaq II 2xGreen Master Mix.

PCR products were separated electrophoretically by transferring 5 μL of every amplified DNA fragment to 1.5% agarose gel stained with ethidium bromide (0.5 μg/mL) (Sigma, St. Louis, MO, USA). Electrophoresis was conducted for 1 h at 100 V in 0.5× TBE buffer.

Table 2. Polymerase chain reaction (PCR) primers and parameters.

Target Gene	Primer Sequence (5′-3′)	Amplicon Size (bp)	PCR Annealing Temp (°C)	References
*sul*1	CGCACCGGAAACATCGCTGCAC	163	55.9	[41]
	TGAAGTTCCGCCGCAAGGCTCG			
*sul*2	TCCGGTGGAGGCCGGTATCTGG	191	60.8	
	CGGGAATGCCATCTGCCTTGAG			
*qep*A	CCAGCTCGGCAACTTGATAC	570	58	[42]
	ATGCTCGCCTTCCAGAAAA			
aac(6′)-Ib-cr	TTGCGATGCTCTATGAGTGGCTA	482	55	[43]
	CTCGAATGCCTGGCGTGTTT			

2.4. Cluster Analysis

An analysis of hierarchical clustering was performed to illustrate the relationship of the sampling sites and the collection season based on the occurrence of studied ARGs. The Dice similarity coefficient (Sørensen–Dice index) was used to measure the relationship between two sets of data. The unweighted pair group method with arithmetic mean (UPGMA) was used as a clustering method. A hierarchical tree (dendrogram) of the analyzed samples was generated for each research object (Supplementary materials). The clustering analysis was performed using the Molecular Evolutionary Genetics Analysis (MEGA7, Pennsylvania State University, State College, PA, USA, 2016) software [44].

3. Results and Discussion

The results of the cluster analysis did not show a similarity between the results for samples collected in the same season. The samples collected in different seasons from the same sampling sites were often grouped in one cluster on the dendrogram. It can be concluded that the sampling season probably had no effect on the presence of ARGs in the analyzed samples (Figure S1). However, the prevalence of ARGs differed across the types of genes and the stages of wastewater treatment. Raw sewage was a significant source of ARGs and contained all of the analyzed genes (Table 3).

Influent wastewater is an important reservoir of ARGs. Wastewater treatment plants offer supportive conditions for the spread of ARGs by horizontal gene transfer (HGT) [45–47]. The low effectiveness of treatment processes can contribute to the transmission of ARGs from treated effluents to surface water [2,19,48]. Antibiotic resistance genes are localized on MGEs, which can be exchanged by both closely related and phylogenetically distant bacterial species. As a result, antibiotic resistance can be spread between bacteria of anthropogenic origin and bacterial communities in the natural environment [6,42]. ARGs in the environment stays in two forms: intracellular ARGs (iARGs) and extracellular ARGs (eARGs). The forms of ARGs that occur in the environment determine the character of its further transmission. iARGs support the spreading of antibiotic resistance via conjugation and transduction, while eARGs can be uptaken and integrated by the competent non-resistant bacteria via natural transformation. Of the mentioned HGT mechanisms, conjugation is deemed to have the biggest impact on the dissemination of ARGs, while transformation and transduction are considered less important [49].

Sulfonamides were the first safe and effective antimicrobial drugs in clinical practice that targeted selected bacteria [50]. The medical use of sulfonamides is decreasing in Europe. The compound annual growth rate (CAGR) of the sulfonamide market in Europe reached negative values between 2009 and 2018 [36]. In 2017, sulfonamides were the least used class of antibiotics for systemic use in Polish and European hospitals [51]. Sulfonamide resistance in Gram-negative bacteria can probably be attributed to *sul*1 and *sul*2 genes, which are carried by plasmids. Research has demonstrated that sulfonamide resistance genes are the most ubiquitous ARGs in the environment [19,39]. In the present study, the *sul*1 gene was identified in all wastewater and sewage sludge samples collected in both WWTPs. This gene was also present in samples of river water collected upstream from the effluent discharge point. The above findings indicate that the *sul*1 gene is widely spread in the environment. The analyzed gene was not eliminated in successive stages of treatment, which may increase the concentration of this gene in the water and bottom sediments of effluent-receiving rivers. Surface waters are highly contaminated with the *sul*1 gene [15,16,18,52]. The *sul*1 gene has been also identified in samples of municipal wastewater [53], activated sludge [47] and wastewater containing animal waste [54]. Hospital sewage appears to be the major source of the *sul*1 gene in the environment. According to Lye et al. [2], the prevalence of *sul*1 is considerably higher in hospital wastewater than in other types of sewage. Similar results were reported by Wang et al. [14], who analyzed hospital sewage in China. In other studies, hospital wastewater was characterized by the highest concentration of the *sul*1 gene relative to other ARGs [13,55]. The ubiquitous character of *sul*1 could result from a close relationship between *sul*1 and class 1 integrons that are widespread in the environment [56]. According to Poey et al. [57], *sul*1 occurs in the variable region (gene cassette) of class 1 integrons. In the current study, hospital sewage accounted for 2–6% of the wastewater processed by the WWTPs. WM-WWTP treats wastewater from three hospitals, and S-WWTP treats wastewater from one hospital. Hospital sewage could be a major source of *sul*1 in the analyzed WWTPs, which could explain the high prevalence of this gene in all samples.

Table 3. Presence of amplicons for genes encoding resistance to sulfonamides and fluoroquinolones in the analyzed samples. Colors correspond to sampling sites as in Table 1. + and – denote amplicons for each gene that were and were not detected via endpoint PCR, respectively.

	Symbol	*sul*1		*sul*2		*qepA*		*aac(6′)-Ib-cr*	
		Summer	Fall	Summer	Fall	Summer	Fall	Summer	Fall
S-WWTP *	S1	+	+	+	+	+	+	+	+
	S2	+	+	+	+	–	+	+	+
	S3	+	+	+	–	–	+	+	+
	S4	+	+	+	–	–	–	+	+
	S5	+	+	+	–	–	+	+	+
	S6	+	+	–	+	–	–	+	+
	S7	+	+	+	+	+	+	+	+
	S8	+	+	–	+	–	–	+	+
	S9	+	+	–	+	–	–	+	+
WM-WWTP **	W1	+	+	+	+	+	+	+	+
	W2	+	+	+	+	+	+	+	+
	W3	+	+	+	+	–	+	+	+
	W4	+	+	+	+	+	–	+	+
	W5	+	+	+	–	+	+	+	+
	W6	+	+	+	+	–	–	+	+
	W7	+	+	+	+	+	–	+	+
	W8	+	+	–	–	–	–	+	+
	W9	+	+	+	–	–	+	+	+

* Silesian wastewater treatment plant; ** Warmia and Mazury wastewater treatment plant.

The *sul*2 gene occurred less frequently than the *sul*1 gene. The *sul*2 gene was not identified in river water upstream from the effluent discharge point (excluding site S8 in fall). This gene was present in all samples of raw wastewater, but it was eliminated in successive stages of treatment. These observations could also suggest that *sul*2 was less abundant in raw sewage that *sul*1. The presence of the *sul*2 gene in river water was noted only in WM-WWTP in summer. Koczura et al. [52] reported significantly smaller concentrations of *sul*2 than *sul*1 in surface waters. In a study by Ziembińska-Buczyńska et al. [47], *sul*2 was less prevalent than *sul*1 in bacterial isolates from activated sludge. According to Pei et al. [41], *sul*2 is an important indicator of the environmental impacts of agriculture, which could explain why *sul*2 concentrations are higher in areas with a predominance of agriculture, in particular livestock production and aquaculture. Lye et al. [2] found the highest abundance of *sul*2 in zoo wastewater. The livestock population in Warmia and Mazury is several times higher than in Silesia [58], which could imply that Warmia and Mazury is characterized by a higher contamination of animal waste and higher concentrations of *sul*2 in the environment, in particular in wastewater evacuated from animal farms. Olsztyn, the capital city of Warmia and Mazury, where MW-WWTP is situated, is the seat of Poland's largest poultry company, which specializes in turkey rearing and the production and processing of turkey meat. Wastewater from turkey farms and production plants is evacuated to MW-WWTP, which could explain the higher abundance of *sul*2 in raw sewage reaching MW-WWTP than S-WWTP. The *sul*2 gene was not identified in only one stage of wastewater treatment in MW-WWTP, whereas in S-WWTP, *sul*2 was not detected in four sampling sites.

According to the literature, seasonal variations in the virulence of pathogenic microorganisms and the immune status of infected individuals are responsible for the seasonality of infectious diseases [59]. Seasonal variations are also noted in antibiotic use, and the consumption of antimicrobials is significantly higher in fall than in summer [60]. Despite the above, it appears that the prevalence of sulfonamide resistance genes in samples of river water, wastewater and sludge was not noticeably affected by season in the present study; however, this is only a preliminary case study and to determine more precise correlations the research must be expanded with quantitative analyses of ARGs. Similar observations were made by Koczura et al. [52], who did not report significant seasonal differences in *sul*1 copy numbers in river water and sewage sludge, but noted that the *sul*2 gene was more prevalent in spring samples.

Fluoroquinolones are broad-spectrum antibiotics and one of the most frequently prescribed antimicrobials. Extensive clinical use of fluoroquinolones has contributed to high resistance of pathogenic microorganisms to this group of antibiotics. In Europe, the number of hospital strains of *Escherchia coli* and *Klebsiella pneumoniae* resistant to fluoroquinolones increased in 2015 –2018. In 2018, the highest percentage of *Pseudomonas aeruginosa* isolates (19.7%) were resistant to fluoroquinolones [33]. The genes conditioning bacterial resistance to fluoroquinolones include *aac(6′)-Ib-cr,* which encodes aminoglycoside acetyltransferase, and *qep*A, which encodes active efflux pumps [34]. In this study, both genes were detected in samples of raw wastewater.

The *aac(6′)-Ib-cr* gene was identified in all analyzed samples, and it was not eliminated during wastewater treatment. It was also detected in samples of river water collected upstream from effluent discharge points, which suggests that this gene also originates from other sources. The *qep*A gene was less frequently isolated, and it was not found in river water sampled upstream from effluent discharge points. In S-WWTP, the *qep*A gene was completely eliminated from treated wastewater in summer, and it was not transmitted to the river with the evacuated effluents. This gene was present only in sewage sludge where the concentration of biological material was highest relative to the remaining samples. The *qep*A gene was less effectively removed in WM-WWTP. In both seasons, *qep*A was not detected in river water sampled upstream from the effluent discharge point, but it was identified in the samples collected downstream from the effluent discharge point.

According to the literature, *aac(6′)-Ib-cr* and *qepA* genes are commonly found in the wastewater. Korzeniewska and Harnisz [21] detected both genes in influents and effluents from 13 wastewater treatment plants with different type of the wastewater treatment technologies modifications. Yan et al. [61] found *aac(6′)-Ib-cr* and *qep*A in treated wastewater from three Chinese hospitals, in municipal wastewater and in river water. In each sampling site, the *aac(6′)-Ib-cr* gene had higher copy numbers than the *qep*A gene. Wen et al. [48] analyzed samples of unpolluted river water and samples of river water collected in the vicinity of hospitals. The *qep*A gene was not detected in any of the samples, whereas the *aac(6′)-Ib-cr* gene was ubiquitous in all samples.

The lower prevalence of *qep*A could be attributed to the fact that resistance to fluoroquinolones encoded by this gene evolved relatively late. The *qep*A gene was first identified in 2007 by research teams from Belgium and Japan [62,63]. The first mechanism of plasmid encoded resistance to fluoroquinolones was discovered in 1998 [64].

It should be noted that all studied genes (excluding *qepA* at site W7 in fall) were present in treated sewage sludge intended for further use. More than 500,000 tons of sewage sludge are produced each year in Poland, of which around 20% are used as agricultural fertilizers on account of their high organic matter content [65]. The application of sewage sludge as agricultural fertilizer contributes to the spread of antibiotic resistance in the environment [28]. Chen et al. [28] found that sewage sludge fertilization can transfer up to 108 of ARGs and MGEs to soil. The resulting increase in bacterial diversity in the soil environment was significantly correlated with the prevalence of ARGs. Sewage sludge is particularly abundant in tetracycline and sulfonamide resistance genes [27]. According to Lee et al. [27], sulfonamide resistance genes (*sul1* and *sul2*) were the most frequently occurring ARGs (45.6%) in the examined sewage sludge. The number of *sul1* copies in sewage sludge significantly exceeded (26–87 times) the number of *sul2* copies. Lee et al. [27] also observed that the prevalence of sulfonamide resistance genes was six-fold (%) higher in sewage sludge than in raw wastewater, which suggests that these genes are highly accumulated in sewage sludge. Quinolone resistance genes were identified significantly less frequently, and they were not present in sewage sludge. The accumulation of ARGs and MGEs in soil could imply that the application of sewage sludge could accelerate gene transmission to the soil environment via HGT.

The presence of antibiotics acting as active mutagens can also significantly contribute to the evolution of drug resistance mechanisms. Sublethal concentrations of antibiotics, which are frequently noted in wastewater, can lead to mutations that promote the emergence of drug resistance [66,67]. The samples of raw sewage, treated wastewater and river water that were examined in the current study were additionally analyzed by Giebułtowicz et al. [68] for the presence of 26 antimicrobials. Very high concentrations of sulfamethoxazole (SXT) and ciprofloxacin (CIP) were noted in raw wastewater reaching both WWTPs, and they were nearly three times higher in WM-WWTP than in S-WWTP. It should also be noted that CIP levels in raw wastewater exceeded the minimum inhibitory concentrations (MIC) recommended by EUCAST for most microorganisms [69]. To regulate the risk issuing from the concentrations of antibiotics in the environment, Bengtson-Palme and Larsson [70] estimated Predicted No-Effect Concentrations (PENCs) of antimicrobials for resistance selection. Based on these [70] calculations, average CIP concentrations in influents, effluents and downstream rivers were found to exceed the PENCs values for influents, effluents and downstream rivers in both WWTPs, which is particularly worrying. Based on the calculated values of the risk quotient (RQ), Giebułtowicz et al. [68] concluded that high concentrations of the analyzed antibiotics contribute to the development of selective resistance. The results of the qualitative analyses presented in this study cannot reveal a correlation between antibiotic concentrations and the prevalence of the examined ARGs. Further research involving quantitative analysis is needed to investigate the above problem.

The study determined whether the tested ARGs are present/absent in the DNA isolated from analyzed samples. However, it remains unknown if and what part of the current ARGs is intercellularly carried (iARGs) by live and metabolically active bacterial cells. Therefore, to reach beyond the qualitative and quantitative analyzes of ARGs and to reach a more extensive knowledge of ARGs dynamics, it is necessary to identify which of the detected ARGs are expressed. The combination of metagenomic and metatranscriptomic analyses will allow the identification of ARGs that are not only present but also actively transcribed. There is little work on antibiotic resistance gene expression in WWTP [71–73]. The researchers report that about 65.8% identified ARGs shows transcriptional activity [73], and that there is a significant overexpression of ARGs occurring in an environment that is heavily affected by antibiotic use [71]. We believe that this issue should be the direction of further research in the WWTPs.

4. Conclusions

Wastewater treated at WWTPs is a significant point source of ARGs. The existing wastewater treatment methods do not effectively eliminate ARGs whose concentrations in treated effluents are not routinely monitored. This imperfect process promotes the release of ARGs into the environment.

This study demonstrated that genes encoding resistance to sulfonamides and fluoroquinolones are widespread in the environment and that WWTPs contribute to their transmission. The prevalence of ARGs in raw wastewater and in various stages of wastewater treatment differed in the examined WWTPs. These variations could be attributed to differences in the type and sources of processed wastewater. WM-WWTP processes wastewater from livestock farms, which could explain the higher prevalence of the *sul*2 gene in the analyzed samples. Hospital wastewater appears to be the main source of *sul*1 in both WWTPs. Sewage sludge was found to be a significant reservoir of ARGs. Sewage sludge should be stabilized before it is used as fertilizer, and its application in agriculture should be monitored. The growing prevalence and spread of ARB and ARGs pose a significant public health concern around the world. A sound knowledge of the sources and transmission mechanisms of antibiotic resistance in various environments is required to develop effective strategies for managing these risks and evaluating their impact on human health. The study is a preliminary study and a base for further in-depth metagenomic and metatranscriptomic analyses.

Supplementary Materials: The following are available online at http://www.mdpi.com/2076-3417/10/17/5816/s1, Figure S1: Similarity of samples, based on the occurrence of antibiotic-resistance genes (ARGs).

Author Contributions: Conceptualization, M.H., E.K. and G.P.; methodology, M.H. and E.K., software, D.R.; validation, M.H. and D.R.; formal analysis, D.R.; resources, M.H., Ł.J. and G.P.; data curation, D.R.; writing—original draft preparation, D.R.; writing—review and editing, M.H., E.K. and G.P.; visualization, D.R.; supervision, M.H.; funding acquisition, M.H. and G.P. All authors have read and agreed to the published version of the manuscript.

Funding: This research was funded by grants from the National Science Center (Poland) No. 2017/26/M/NZ9/00071 and 2017/27/B/NZ9/00267.

Conflicts of Interest: The authors declare no conflict of interest.

References

1. Guo, J.; Li, J.; Chen, H.; Bond, P.L.; Yuan, Z. Metagenomic analysis reveals wastewater treatment plants as hotspots of antibiotic resistance genes and mobile genetic elements. *Water Res.* **2017**, *123*, 468–478. [CrossRef]
2. Lye, Y.L.; Bong, C.W.; Lee, C.W.; Zhang, R.J.; Zhang, G.; Suzuki, S.; Chai, L.C. Anthropogenic impacts on sulfonamide residues and sulfonamide resistant bacteria and genes in Larut and Sangga Besar River, Perak. *Sci. Total Environ.* **2019**, *688*, 1335–1347. [CrossRef]
3. WHO. *WHO Report on Surveillance of Antibiotic Consumption: 2016–2018 Early Implementation*; Licence: CC BY-NC-SA 3.0 IGO; World Health Organization: Geneva, Switzerland, 2018.
4. Danner, M.C.; Robertson, A.; Behrends, V.; Reiss, J. Antibiotic pollution in surface fresh waters: Occurrence and effects. *Sci. Total Environ.* **2019**, *664*, 793–804. [CrossRef]
5. Koniuszewska, I.; Korzeniewska, E.; Harnisz, M.; Kiedrzyńska, E.; Kiedrzyński, M.; Czatzkowska, M.; Jarosiewicz, P.; Zalewski, M. The occurrence of antibiotic-resistance genes in the Pilica River, Poland. *Ecohydrol. Hydrobiol.* **2020**, *20*, 1–11. [CrossRef]
6. Xiong, W.; Sun, Y.; Ding, X.; Zhang, Y.; Zeng, Z. Antibiotic resistance genes occurrence and bacterial community composition in the Liuxi River. *Front. Environ. Sci.* **2014**, *2*, 1–6. [CrossRef]
7. Andrade, L.; Kelly, M.; Hynds, P.; Weatherill, J.; Majury, A.; O'Dwyer, J. Groundwater resources as a global reservoir for antimicrobial-resistant bacteria. *Water Res.* **2020**, *170*, 115360. [CrossRef]
8. Burke, V.; Richter, D.; Greskowiak, J.; Mehrtens, A.; Schulz, L.; Massmann, G.; Victoria, B.; Doreen, R.; Janek, G.; Anne, M.; et al. Occurrence of Antibiotics in Surface and Groundwater of a Drinking Water Catchment Area in Germany. *Water Environ. Res.* **2016**, *88*, 652–659. [CrossRef]
9. Szekeres, E.; Chiriac, C.M.; Baricz, A.; Szőke-Nagy, T.; Lung, I.; Soran, M.-L.; Rudi, K.; Dragoş, N.; Coman, C. Investigating antibiotics, antibiotic resistance genes, and microbial contaminants in groundwater in relation to the proximity of urban areas. *Environ. Pollut.* **2018**, *236*, 734–744. [CrossRef] [PubMed]
10. Pérez-Valera, E.; Kyselková, M.; Ahmed, E.; Sladecek, F.X.J.; Goberna, M.; Elhottová, D. Native soil microorganisms hinder the soil enrichment with antibiotic resistance genes following manure applications. *Sci. Rep.* **2019**, *9*, 6760. [CrossRef] [PubMed]
11. Forsberg, K.J.; Patel, S.; Gibson, M.K.; Lauber, C.L.; Knight, R.; Fierer, N.; Dantas, G. Bacterial phylogeny structures soil resistomes across habitats. *Nature* **2014**, *509*, 612–616. [CrossRef] [PubMed]

12. Cycoń, M.; Mrozik, A.; Piotrowska-Seget, Z. Antibiotics in the Soil Environment—Degradation and Their Impact on Microbial Activity and Diversity. *Front. Microbiol.* **2019**, *10*, 338. [CrossRef]
13. Lorenzo, P.; Adriana, A.; Jéssica, S.; Carles, B.; Marinella, F.; Marta, L.; Luis, B.J.; Pierre, S. Antibiotic resistance in urban and hospital wastewaters and their impact on a receiving freshwater ecosystem. *Chemosphere* **2018**, *206*, 70–82. [CrossRef] [PubMed]
14. Wang, Q.; Wang, P.; Yang, Q. Occurrence and diversity of antibiotic resistance in untreated hospital wastewater. *Sci. Total Environ.* **2018**, *621*, 990–999. [CrossRef] [PubMed]
15. Cacace, D.; Fatta-Kassinos, D.; Manaia, C.M.; Cytryn, E.; Kreuzinger, N.; Rizzo, L.; Karaolia, P.; Schwartz, T.; Alexander, J.; Merlin, C.; et al. Antibiotic resistance genes in treated wastewater and in the receiving water bodies: A pan-European survey of urban settings. *Water Res.* **2019**, *162*, 320–330. [CrossRef] [PubMed]
16. Huang, Z.; Zhao, W.; Xu, T.; Zheng, B.; Yin, D. Occurrence and distribution of antibiotic resistance genes in the water and sediments of Qingcaosha Reservoir, Shanghai, China. *Environ. Sci. Eur.* **2019**, *31*, 1–9. [CrossRef]
17. Osińska, A.; Korzeniewska, E.; Harnisz, M.; Felis, E.; Bajkacz, S.; Jachimowicz, P.; Niestępski, S.; Konopka, I. Small-scale wastewater treatment plants as a source of the dissemination of antibiotic resistance genes in the aquatic environment. *J. Hazard. Mater.* **2020**, *381*, 121221. [CrossRef]
18. Sabri, N.A.; Schmitt, H.; Van Der Zaan, B.; Gerritsen, H.W.; Zuidema, T.; Rijnaarts, H.H.M.; Langenhoff, A.A.M. Prevalence of antibiotics and antibiotic resistance genes in a wastewater effluent-receiving river in the Netherlands. *J. Environ. Chem. Eng.* **2020**, *8*, 102245. [CrossRef]
19. Stange, C.; Yin, D.; Xu, T.; Guo, X.; Schäfer, C.; Tiehm, A. Distribution of clinically relevant antibiotic resistance genes in Lake Tai, China. *Sci. Total Environ.* **2019**, *655*, 337–346. [CrossRef]
20. Rizzo, L.; Manaia, C.M.; Merlin, C.; Schwartz, T.; Dagot, C.; Ploy, M.C.; Michael, I.; Fatta-Kassinos, D. Urban wastewater treatment plants as hotspots for antibiotic resistant bacteria and genes spread into the environment: A review. *Sci. Total Environ.* **2013**, *447*, 345–360. [CrossRef]
21. Korzeniewska, E.; Harnisz, M. Relationship between modification of activated sludge wastewater treatment and changes in antibiotic resistance of bacteria. *Sci. Total Environ.* **2018**, *639*, 304–315. [CrossRef]
22. Pazda, M.; Kumirska, J.; Stepnowski, P.; Mulkiewicz, E. Antibiotic resistance genes identified in wastewater treatment plant systems—A review. *Sci. Total Environ.* **2019**, *697*, 134023. [CrossRef] [PubMed]
23. Jäger, T.; Hembach, N.; Elpers, C.; Wieland, A.; Alexander, J.; Hiller, C.; Krauter, G.; Schwartz, T. Reduction of Antibiotic Resistant Bacteria During Conventional and Advanced Wastewater Treatment, and the Disseminated Loads Released to the Environment. *Front. Microbiol.* **2018**, *9*, 2599. [CrossRef] [PubMed]
24. Marano, R.B.M.; Cytryn, E. The Mobile Resistome in Wastewater Treatment Facilities and Downstream Environments. In *Antimicrobial Resistance in Wastewater Treatment Processes*; Keen, P.L., Fugère, R., Eds.; John Wiley & Sons, Inc. (Wiley): Hoboken, NJ, USA, 2017; pp. 129–155.
25. Kaplan, E.; Ofek, M.; Jurkevitch, E.; Cytryn, E. Characterization of fluoroquinolone resistance and qnr diversity in Enterobacteriaceae from municipal biosolids. *Front. Microbiol.* **2013**, *4*, 144. [CrossRef] [PubMed]
26. Minister of Economy. Internet System of Legal Acts. Available online: http://isap.sejm.gov.pl/isap.nsf/DocDetails.xsp?id=WDU20150001277 (accessed on 21 August 2020). (In Polish)
27. Lee, J.; Shin, S.G.; Jang, H.M.; Kim, Y.B.; Lee, J.; Kim, Y.M. Characterization of antibiotic resistance genes in representative organic solid wastes: Food waste-recycling wastewater, manure, and sewage sludge. *Sci. Total Environ.* **2017**, *579*, 1692–1698. [CrossRef]
28. Chen, Q.-L.; An, X.; Li, H.; Su, J.; Ma, Y.; Zhu, Y.-G. Long-term field application of sewage sludge increases the abundance of antibiotic resistance genes in soil. *Environ. Int.* **2016**, *92*, 1–10. [CrossRef]
29. Gao, P.; Munir, M.; Xagoraraki, I. Correlation of tetracycline and sulfonamide antibiotics with corresponding resistance genes and resistant bacteria in a conventional municipal wastewater treatment plant. *Sci. Total Environ.* **2012**, *421*, 173–183. [CrossRef]
30. Munir, M.; Xagoraraki, I. Levels of Antibiotic Resistance Genes in Manure, Biosolids, and Fertilized Soil. *J. Environ. Qual.* **2011**, *40*, 248–255. [CrossRef]
31. Ma, Y.; Wilson, C.A.; Novak, J.T.; Riffat, R.; Aynur, S.; Murthy, S.; Pruden, A. Effect of Various Sludge Digestion Conditions on Sulfonamide, Macrolide, and Tetracycline Resistance Genes and Class I Integrons. *Environ. Sci. Technol.* **2011**, *45*, 7855–7861. [CrossRef]

32. Yang, L.; Liu, W.; Zhu, D.; Hou, J.; Ma, T.; Wu, L.; Zhu, Y.; Christie, P. Application of biosolids drives the diversity of antibiotic resistance genes in soil and lettuce at harvest. *Soil Biol. Biochem.* **2018**, *122*, 131–140. [CrossRef]
33. European Centre for Disease Prevention and Control (ECDC). *Surveillance of Antimicrobial Resistance in Europe 2018*; Annual report of the European Antimicrobial Resistance Surveillance Network (EARS-Net); ECDC: Stockholm, Sweden, 2018.
34. Ruiz, J.; Pons, M.J.; Gomes, C. Transferable mechanisms of quinolone resistance. *Int. J. Antimicrob. Agents* **2012**, *40*, 196–203. [CrossRef]
35. Wong, M.H.; Chan, E.W.; Liu, L.Z.; Chen, S. PMQR genes oqxAB and aac(6′)Ib-cr accelerate the development of fluoroquinolone resistance in Salmonella typhimurium. *Front. Microbiol.* **2014**, *5*, 521. [CrossRef] [PubMed]
36. European Centre for Disease Prevention and Control (ECDC). *Antimicrobial Consumption—Annual Epidemiological Report for 2018*; ECDC: Stockholm, Sweden, 2019.
37. Jiang, H.; Cheng, H.; Liang, Y.; Yu, S.; Yu, T.; Fang, J.; Zhu, C. Diverse Mobile Genetic Elements and Conjugal Transferability of Sulfonamide Resistance Genes (sul1, sul2, and sul3) in Escherichia coli Isolates From Penaeus vannamei and Pork From Large Markets in Zhejiang, China. *Front. Microbiol.* **2019**, *10*, 1787. [CrossRef]
38. Hsu, J.T.; Chen, C.Y.; Young, C.W.; Chao, W.L.; Li, M.H.; Liu, Y.H.; Lin, C.M.; Ying, C. Prevalence of sulfonamide-resistant bacteria, resistance genes and integron-associated horizontal gene transfer in natural water bodies and soils adjacent to a swine feedlot in northern Taiwan. *J. Hazard. Mater.* **2014**, *277*, 34–43. [CrossRef] [PubMed]
39. Yan, M.; Xu, C.; Huang, Y.; Nie, H.; Wang, J. Tetracyclines, sulfonamides and quinolones and their corresponding resistance genes in the Three Gorges Reservoir, China. *Sci. Total Environ.* **2018**, 840–848. [CrossRef] [PubMed]
40. Romańczuk, P.; Pieczykolan, M. Porównanie eksploatowanych ciągów technologicznych ściekowych na oczyszczalni ścieków Tychy–Urbanowice. *Gaz Woda Tech. Sanit.* **2006**, *7-8*, 49–53.
41. Pei, R.; Kim, S.C.; Carlson, K.H.; Pruden, A. Effect of River Landscape on the sediment concentrations of antibiotics and corresponding antibiotic resistance genes (ARG). *Water Res.* **2006**, *40*, 2427–2435. [CrossRef]
42. Li, J.; Wang, T.; Shao, B.; Shen, J.; Wang, S.; Wu, Y. Plasmid-Mediated Quinolone Resistance Genes and Antibiotic Residues in Wastewater and Soil Adjacent to Swine Feedlots: Potential Transfer to Agricultural Lands. *Environ. Heal. Perspect.* **2012**, *120*, 1144–1149. [CrossRef]
43. Park, C.H.; Robicsek, A.; Jacoby, G.A.; Sahm, D.; Hooper, D.C. Prevalence in the United States of aac(6′)-Ib-cr Encoding a Ciprofloxacin-Modifying Enzyme. *Antimicrob. Agents Chemother.* **2006**, *50*, 3953–3955. [CrossRef]
44. Kumar, S.; Stecher, G.; Tamura, K. MEGA7: Molecular Evolutionary Genetics Analysis version 7.0 for bigger datasets. *Mol. Boil. Evol.* **2016**, *33*, 1870–1874. [CrossRef]
45. Kurasam, J.; Sihag, P.; Mandal, P.K.; Sarkar, S. Presence of fluoroquinolone resistance with persistent occurrence of gyrA gene mutations in a municipal wastewater treatment plant in India. *Chemosphere* **2018**, *211*, 817–825. [CrossRef]
46. Osińska, A.; Harnisz, M.; Korzeniewska, E. Prevalence of plasmid-mediated multidrug resistance determinants in fluoroquinolone-resistant bacteria isolated from sewage and surface water. *Environ. Sci. Pollut. Res.* **2016**, *23*, 10818–10831. [CrossRef]
47. Ziembińska-Buczyńska, A.; Felis, E.; Folkert, J.; Meresta, A.; Stawicka, D.; Gnida, A.; Surmacz-Górska, J. Detection of antibiotic resistance genes in wastewater treatment plant—Molecular and classical approach. *Arch. Environ. Prot.* **2015**, *41*, 23–32. [CrossRef]
48. Wen, Y.; Pu, X.; Zheng, W.; Hu, G. High Prevalence of Plasmid-Mediated Quinolone Resistance and IncQ Plasmids Carrying qnrS2 Gene in Bacteria from Rivers near Hospitals and Aquaculture in China. *PLoS ONE* **2016**, *11*, e0159418. [CrossRef] [PubMed]
49. Dong, P.; Wang, H.; Fang, T.; Wang, Y.; Ye, Q. Assessment of extracellular antibiotic resistance genes (eARGs) in typical environmental samples and the transforming ability of eARG. *Environ. Int.* **2019**, *125*, 90–96. [CrossRef] [PubMed]
50. Tačić, A.; Nikolić, V.; Nikolić, L.; Savić, I. Antimicrobial sulfonamide drugs. *Adv. Technol.* **2017**, *6*, 58–71. [CrossRef]
51. Gouvras, G. The European Centre for Disease Prevention and Control. *Eurosurveillance* **2004**, *9*, 2. [CrossRef]

52. Koczura, R.; Mokracka, J.; Taraszewska, A.; Łopacinska, N. Abundance of Class 1 Integron-Integrase and Sulfonamide Resistance Genes in River Water and Sediment is Affected by Anthropogenic Pressure and Environmental Factors. *Microb. Ecol.* **2016**, *72*, 909–916. [CrossRef]
53. Mohameda, H.S.A.; Uswege, M.; Robinson, H.M. Correlation between antibiotic concentrations genes contaminations at Mafisa wastewater treatment plant in Morogoro Municipality, Tanzania. *Glob. Environ. Health Saf.* **2018**, *2*, 5.
54. Tao, C.-W.; Hsu, B.-M.; Ji, W.-T.; Hsu, T.-K.; Kao, P.-M.; Hsu, C.-P.; Shen, S.-M.; Shen, T.-Y.; Wan, T.-J.; Huang, Y.-L. Evaluation of five antibiotic resistance genes in wastewater treatment systems of swine farms by real-time PCR. *Sci. Total Environ.* **2014**, *496*, 116–121. [CrossRef]
55. Szekeres, E.; Baricz, A.; Chiriac, C.M.; Farkas, A.; Opriş, O.; Soran, M.-L.; Andrei, A.Ş.; Rudi, K.; Balcázar, J.L.; Dragos, N.; et al. Abundance of antibiotics, antibiotic resistance genes and bacterial community composition in wastewater effluents from different Romanian hospitals. *Environ. Pollut.* **2017**, *225*, 304–315. [CrossRef]
56. Gundogdu, A.; Long, Y.B.; Vollmerhausen, T.L.; Katouli, M. Antimicrobial resistance and distribution of sul genes and integron-associated intI genes among uropathogenic Escherichia coli in Queensland, Australia. *J. Med. Microbiol.* **2011**, *60*, 1633–1642. [CrossRef] [PubMed]
57. Poey, M.E.; Azpiroz, M.F.; Laviña, M. On sulfonamide resistance, sul genes, class 1 integrons and their horizontal transfer in Escherichia coli. *Microb. Pathog.* **2019**, *135*, 103611. [CrossRef] [PubMed]
58. Statistics Poland; Agriculture Department. Farm Animals in 2017. 2018. Available online: https://stat.gov.pl/files/gfx/portalinformacyjny/pl/defaultaktualnosci/5508/6/18/1/zwierzeta_gospodarskie_w_2017_roku.pdf (accessed on 2 July 2020).
59. Fares, A. Factors Influencing the Seasonal Patterns of Infectious Diseases. *Int. J. Prev. Med.* **2013**, *4*, 128–132. [PubMed]
60. Zachara, J.; Zachara, R.; Zachara, N.; Matuszczak, A.; Kłoda, K. The comparison of use of antibiotics due to acute respiratory infections in the rural population of primary care in 2010 and 2017. *J. Educ. Health Sport* **2019**, *9*, 70–83.
61. Yan, L.; Liu, D.; Wang, X.-H.; Wang, Y.; Zhang, B.; Wang, M.; Xu, H. Bacterial plasmid-mediated quinolone resistance genes in aquatic environments in China. *Sci. Rep.* **2017**, *7*, 40610. [CrossRef] [PubMed]
62. Périchon, B.; Courvalin, P.; Galimand, M. Transferable Resistance to Aminoglycosides by Methylation of G1405 in 16S rRNA and to Hydrophilic Fluoroquinolones by QepA-Mediated Efflux in Escherichia coli. *Antimicrob. Agents Chemother.* **2007**, *51*, 2464–2469. [CrossRef] [PubMed]
63. Yamane, K.; Wachino, J.-I.; Suzuki, S.; Kimura, K.; Shibata, N.; Kato, H.; Shibayama, K.; Konda, T.; Arakawa, Y. New Plasmid-Mediated Fluoroquinolone Efflux Pump, QepA, Found in an Escherichia coli Clinical Isolate. *Antimicrob. Agents Chemother.* **2007**, *51*, 3354–3360. [CrossRef]
64. Martínez-Martínez, L.; Pascual, A.; Jacoby, G.A. Quinolone resistance from a transferable plasmid. *Lancet* **1998**, *351*, 797–799. [CrossRef]
65. Eurostat. Sewage Sludge Production and Disposal from Urban Wastewater (in Dry Substance (d.s)). Dataset [Tables], Product Code: Ten00030, Updated on 31 Jan 2020. Available online: http://ec.europa.eu/eurostat/tgm/table.do?tab=table&plugin=1&language=en&pcode=ten00030 (accessed on 2 July 2020).
66. Nair, C.G.; Chao, C.; Ryall, B.; Williams, H.D. Sub-lethal concentrations of antibiotics increase mutation frequency in the cystic fibrosis pathogen Pseudomonas aeruginosa. *Lett. Appl. Microbiol.* **2013**, *56*, 149–154. [CrossRef]
67. Kohanski, M.A.; Depristo, M.A.; Collins, J.J. Sublethal Antibiotic Treatment Leads to Multidrug Resistance via Radical-Induced Mutagenesis. *Mol. Cell* **2010**, *37*, 311–320. [CrossRef]
68. Giebułtowicz, J.; Nałęcz-Jawecki, G.; Harnisz, M.; Kucharski, D.; Korzeniewska, E.; Płaza, G. Environmental Risk and Risk of Resistance Selection Due to Antimicrobials' Occurrence in Two Polish Wastewater Treatment Plants and Receiving Surface Water. *Molecules* **2020**, *25*, 1470. [CrossRef] [PubMed]
69. European Committee on Antimicrobial Susceptibility Testing Breakpoint Tables for Interpretation of MICs and Zone Diameters Version 9.0, Valid from 1 January 2019. Available online: https://korld.nil.gov.pl/pdf/EUCAST_breakpoints_tlumaczenie_wersja%209.0_strona.pdf (accessed on 2 July 2020). (in Polish)
70. Bengtsson-Palme, J.; Larsson, D.G.J. Concentrations of antibiotics predicted to select for resistant bacteria: Proposed limits for environmental regulation. *Environ. Int.* **2016**, *86*, 140–149. [CrossRef]

71. Rowe, W.P.M.; Baker-Austin, C.; Verner-Jeffreys, D.W.; Ryan, J.J.; Micallef, C.; Maskell, D.J.; Pearce, G.P. Overexpression of antibiotic resistance genes in hospital effluents over time. *J. Antimicrob. Chemother.* **2017**, *72*, 1617–1623. [CrossRef] [PubMed]
72. Ju, F.; Beck, K.; Yin, X.; Maccagnan, A.; McArdell, C.S.; Singer, H.P.; Johnson, D.R.; Zhang, T.; Bürgmann, H. Wastewater treatment plant resistomes are shaped by bacterial composition, genetic exchange, and upregulated expression in the effluent microbiomes. *ISME J.* **2019**, *13*, 346–360. [CrossRef] [PubMed]
73. Liu, Z.; Klümper, U.; Liu, Y.; Yang, Y.; Wei, Q.; Lin, J.-G.; Gu, J.-D.; Li, M. Metagenomic and metatranscriptomic analyses reveal activity and hosts of antibiotic resistance genes in activated sludge. *Environ. Int.* **2019**, *129*, 208–220. [CrossRef] [PubMed]

Article

Composition Characteristics of Organic Matter and Bacterial Communities under the *Alternanthera philoxeroide* Invasion in Wetlands

Qingqing Cao [1,2], Haijie Zhang [3], Wen Ma [4], Renqing Wang [2] and Jian Liu [2,*]

1 School of Architecture and Urban Planning, Shandong Jianzhu University, Jinan 250101, China; caoqingqing18@sdjzu.edu.cn
2 Environment Research Institute, Shandong University, Qingdao 266237, China; wrq@sdu.edu.cn
3 State Key Laboratory of Biocontrol, Guangdong Provincial Key Laboratory of Plant Resources, School of Life Sciences, Sun Yat-Sen University, Guangzhou 510275, China; zhanghj53@mail.sysu.edu.cn
4 College of Chemistry and Chemical Engineering, Qilu Normal University, Jinan 250200, China; 20177736@qlnu.edu.cn
* Correspondence: ecology@sdu.edu.cn; Tel.: +86–532-58631969

Received: 29 June 2020; Accepted: 10 August 2020; Published: 12 August 2020

Abstract: The influence of *Alternanthera philoxeroide* (alligator weed) invasion on wetland organic matter (OM) accumulation and bacterial changes is rarely studied, but is possibly an important step for revealing the invasion mechanism. Thus, the distribution characteristics of light fraction organic carbon and nitrogen (LFOC and LFON), and heavy fractions organic carbon and nitrogen (HFOC and HFON) were analyzed. Sampling was done on two sediment depths (0–15 cm and 15–25 cm) of invaded and normal habitats of two natural wetlands and two constructed wetlands, and bacterial taxa and composition in surface sediments were also analyzed by high-throughput sequencing. In the surface sediments, the LFOC and LFON contents were significantly higher in the constructed wetlands (0.791 and 0.043 $g \cdot kg^{-1}$) than in the natural wetlands (0.500 and 0.022 $g \cdot kg^{-1}$), and the contents of the C and N fractions were also prominently higher in the invaded areas than in normal wetland habitats. The OM storage was relatively stable. Proteobacteria (55.94%), Bacteroidetes (5.74%), Acidobacteria (6.66%), and Chloroflexi (4.67%) were the dominant bacterial phyla in the wetlands. The abundance of Acidobacteria, Actinobacteria, and Gemmatimonadetes were significantly higher in the invaded areas than in the normal habitats. The relative high abundance-based coverage estimator (ACE) index in the constructed wetlands and invaded areas suggested the corresponding high bacterial diversity. The significant and positive relationship between Acidobacteria and organic nitrogen concentrations suggested their potential and positive interrelationships. This study demonstrated that the alligator weed invasion could significantly change the compositions of sediment organic matterand bacteria, thus further changing the nutrition cycle and wetland microhabitat.

Keywords: *Alternanthera philoxeroide*; bacterial composition; organic matter; wetland

---

1. Introduction

Exotic plant invasion is a tremendous threat to natural ecosystems, it was also one focus of ecological research over the last few decades [1]. As one kind of water-saturated and multifunctional ecosystem, wetland provides an ideal habitat for aquatic exotic plants to invasion, growth, and reproduction [2,3]. Especially, *Alternanthera philoxeroides* (alligator weed) was a typical invasive plant species in the world, it was first recorded in South America in the 1900s, subsequently spread to North America, and then invaded the Chinese water areas during the 1930s [4]. Its high reproductive and migration ability caused serious damage to local water environment [5], thus alligator weed invasion has attracted great attention. However, previous studies mostly focused on the harm of alligator weed invasion, as well as how to reduce or eliminate it [1,3], while ignoring its effects on the nutrition cycle or microorganism in the wetland ecosystem.

Wetland has a great potential for carbon and nitrogen accumulation, which plays an important role in relieving global warming and purifying the water quality [6,7]. Plenty of studies showed that the exotic plant invasion might affect the accumulation of soil organic carbon (SOC) and organic nitrogen (ON) [8,9]. Soil organic matter (SOM) is the main form of OC and ON, the study on different SOM categories can further reveal the influence of alien plant invasion on the accumulation process of carbon and nitrogen [10]. The SOM can be divided into light fraction organic matter (LFOM) and heavy fraction organic matter (LFOM), through soil density difference [11]. With the soil density less than 1.7 g·cm^{-3}, LFOM is mostly composed of undecomposed or partially decomposed biological residues, and is sensitive to plant types, land-use types, etc. [12]. Thus, it can be an early indicator to evaluate soil OM changes when suffering alligator weed invasion. With a soil density higher than 1.7 g·cm^{-3}, HFOM is relatively stable in a terrestrial ecosystem and can be used to analyze the dynamic changes of stable OM. The influences of several exotic plant invasion on OC and ON accumulation were previously studied [8,13]. For example, the invasion of *Typha* significantly increased the concentrations of SOM, nitrate, and ammonium [8], the *Phragmites australis* invasion also increased the C stock corresponding to the increase in aboveground biomass [9]. While little research involved the effects of alligator weed invasion on SOM.

Alien plant invasion might also change the soil microhabitat through root exudates, and then change the composition of bacterial communities in the soils, which is likely to be the mechanism of plant invasion [14]. Batten et al. [15] reported that both the invasion of *Centaurea solstitialis* and *Aegilops triuncialis* changed the soil bacterial communities, and especially increased the proportions of sulfur-oxidizing bacteria. However, the effects of alligator weed invasion on bacterial composition is still unknown.

Nansi Lake (NL) is one of the largest inland lakes in the South-to-North Water Diversion Project in China. The investigation from 2013–2016 showed that part of the Nansi Lake Basin was invaded by alligator weed, among which constructed wetlands were largely invaded. In addition, previous studies suggested that different wetland types can affect the distribution of LFOM, HFOM, and microorganism [16], which might cause disturbance to the effects of alligator weed invasion. Thus, two natural wetlands and two constructed wetlands from the Nansi Lake Basin were investigated and invaded, and normal wetland sediments were sampled to study the accumulation and composition differences of SOM and bacterial communities under the alligator weed invasion. The aims of this study were—(1) to analyze the content and storage differences of OC and ON by comparing sediments from invaded areas and normal wetland habitat, and (2) to explore and compare the microbial composition under the effects of alligator weed invasion and wetland types. Our results would be important to predict the wetland function under alligator weed invasion.

2. Materials and Methods

2.1. Study Site and Sampling

The field sampling was conducted in the Nansi Lake Basin (NLB), which is located in the Jining, Shandong Province of China. NLB belongs to warm temperate monsoon climate, has a high temperature, is rainy in summer, and sunny and cold weather in winter. The annual average temperature is 13.3–14.1 °C. The average frost-free period is 199 days and the annual precipitation is 597–820 mm.

With a total area of 1266 km^2, the Nansi Lake (NL) is the largest inland lake in the Shandong Province. It has 53 tributaries and the Xinxue River (XR) is a large tributary that connects to NL from the southeast. The Xinxue River Constructed Wetland (XRCW) were designed and operated in 2008, downstream of XR, and were 5100-m-long and 270-m-wide [17]. The Nansi Lake estuary (NLE) is an extensive flat wetland mouth of XRCW that transfers water from the XRCW into NL. NLE has similar characteristic and habitat with XRCW, and also receives the rush from lake water. XRCW and NLE have significantly higher plant richness and diversity than NL and XR. Concretely, the main plants in XRCW and NLE are *P. australis*, *Acorus calamus*, and *A. philoxeroides*, etc., while NL and XR have fewer plant cover. Sediments can be divided into sand, silt, and clay, based on the particle size. In NLB, the corresponding proportions of sand, silt, and clay in NL and XR were on average, 20.21%, 43.30%, and 36.48%, and those in XRCW and NLE were 13.16%, 76.54%, and 10.31%, respectively, based on the previous reports [18,19]. In this study, within the geographic range of 34°42 50.02–34°52 18.51 N and 117°04 45.30–117°18 29.41 E, we collected sediment samples from the XR and NL for characterizing the natural wetlands and from the XRCW and NLE, as representative of constructed wetlands.

The sediment sampling lasted for ten days from May to June of 2017. Sediments of the constructed wetlands were collected first and then the sediments of XR and NL were collected accordingly. Preliminary investigation found that NL and XR were rarely invaded by alligator weed with an invaded area of less than 5%, while XRCW and NLE were largely invaded, with the invaded area exceeding 30%. Thus, the invaded sites and non-invaded sites were selected as representative of invaded habitats and normal habitats in this study. Specifically, we selected eight invaded sites and seven non-invaded sites in XRCW, and three invaded sites and four non-invaded sites in NLE. Two and five non-invaded sites were selected in NL and XR, respectively. In total, 29 sites were selected and the details of the sampling sites are shown in Figure 1. Previous reports showed that OC and ON were sensitive to soil depths in terrestrial ecosystem [20,21], thus sediments from two depths (0–15 cm and 15–25 cm) were collected to analyze the vertical distribution pattern of C and N fractions. At each site, sediment cores of 0–25 cm were excavated, the five-point sampling method was applied to collect sediment samples of 500–1000 g wet weight from two depths, using a GRASP sediment sampler (GRASP ZYQ-WN, Beijing, China). Meanwhile, surface sediment samples of 30–50 g were collected and stored in 4 °C ice box for bacterial analysis, and surface sediments were also in-situ collected using a cutting ring of 100 cm^3 to analyze soil bulk density and moisture contents.

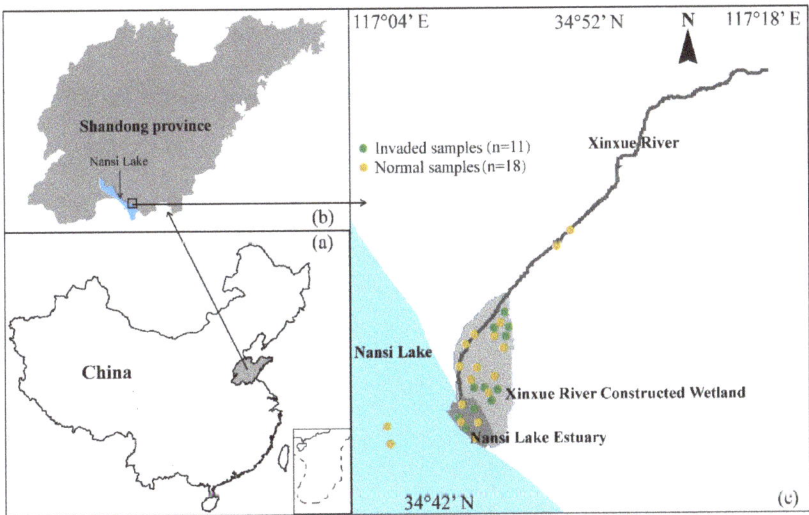

Figure 1. The (**a**) national, (**b**) regional, and (**c**) local geographical setting of the study area. Map (**c**) showed the sampling sites from the invaded and normal habitats of the four wetlands.

2.2. Separation and Analysis of Organic Matter

Sediment samples were air-dried at room temperature (~20 °C) and then ground for further analysis. Due to sensitivity of soil OC and ON analysis, the dried soil was passed through a 0.9-mm sieve to remove biological residues, broken roots, and detritus. The LFOM and HFOM of the weighed 10 g soils were stratified by adding 40 mL 1.70 g·mL^{-1} sodium iodide solution. The C and N contents of the LFOM and HFOM were determined using an elemental analyzer (Vario EL III; Elementar Analysensysteme, Hanau, Germany). From the data thus obtained, we calculated the contents of LFOC, HFOC, LFON, HFON, and the C:N ratios of the light and heavy fractions (LFOC:LFON and HFOC:HFON). Detailed procedures were reported by Zhang et al. [18]. Moisture contents and bulk density were determined and calculated by comparing sediment weight in a set volume (100 cm^3), before and after drying at 105 °C [22]. In addition, the storages of C or N fractions were calculated, based on the sampling depths, bulk density, and C and N fractions contents [23].

2.3. Illumina MiSeq and Data Processing

After the field sampling, the 29 surface samples were freeze-dried for 48 h at −50 °C (F-20; NIHON Co., Shanghai, China) and sifted through a 2-mm nylon sieve to remove residual animal and plant matter, for analyzing the bacterial composition. The cetyltrimethylammonium ammonium bromide was widely used to extract the total bacterial and vegetal genomic DNA in the last decades [24], so it was used in our research. The detailed DNA extracting procedures referred to the report of Niemi et al. [25]. We quantified the extracted DNA by using Nanodrop ND-1000 spectrophotometer (Thermo Fisher Scientific, Waltham, MA, USA) and checked the DNA quality with 1.2% agarose gel electrophoresis [26]. The purified DNA was diluted to 25 ng/μL, and then PCR amplification of the 16S rRNA genes V4 region was performed using the primers 515F and 806R [27]. Sample-specific 7-bp barcodes were incorporated into the primers for multiplex sequencing. The PCR was conducted using the Phusion High-Fidelity PCR Master Mix (Phusion High-Fidelity DNA Polymerase) with GC buffer (New England Biolabs Co., Ipswich, MA, USA) and concrete operations were showed in the report of Dolgova et al. [28]. The PCR products were mixed in equal amounts, according to concentration, and the mixed products were purified with 2% agarose gel electrophoresis and quantified by using the PicoGreen dsDNA Assay Kit (Invitrogen, Carlsbad, CA, USA). Then, the paired-end 2 × 300 bp sequencing was performed using the Illlumina MiSeq platform with MiSeq Reagent Kit v3 at Shanghai Personal Biotechnology Co., Ltd. (Shanghai, China). The Quantitative Insights Into Microbial Ecology (QIIME, v1.8.0) pipeline was employed to process the sequencing data, as previously described [26,27]. The nucleotide sequences we obtained were deposited in the National Center for Biotechnology Information, under the accession number PRJNA470781, PRJNA470783, and PRJNA470794.

Raw sequencing reads with exact matches to the barcodes were assigned to the respective samples and identified as valid sequences in FLASH (version 1.2.7, Center for Computational Biology of Johns Hopkins University, Baltimore, Maryland, USA, 2011. http://ccb.jhu.edu/software/FLASH) [26]. The chimeras were identified and removed by Uchime of Mother (version 4.1, The Department of Microbiology & Immunology at The University of Michigan, Detroit, MI, USA. 2011. http://www.mothur.org/), the low-quality sequences were filtered and the remaining high-quality sequences were clustered into operational taxonomic units (OTUs) at 97% sequence identity by UCLUST, and then representative OTU sequences were selected [16,29]. These sequences were identified at the genus level, against the SILVA rRNA database in Mothur (version 1.31.2, The Department of Microbiology & Immunology at The University of Michigan, Detroit, MI, USA. 2009. http://www.mothur.org/) [30]. On the basis of the taxonomic information obtained, we determined the community composition of each sample at the different classification levels—kingdom, phylum, class, order, family, and genus [31]. To minimize the difference of sequencing depth across samples, an averaged, rounded rarefied OTU table was generated by averaging 100 evenly resampled OTU subsets, under 90% of the minimum sequencing depth.

2.4. Statistical Analysis

Before further statistical analysis, all data were checked by the normality test (Kolmogorov-Smirnov test) and the outliers were removed, based on the corresponding stem-leaf plot. The homogeneity tests of variances were also applied to ensure that the significances were higher than 0.05. The mean concentrations and storages of the C and N fraction were calculated in the four wetlands and one-way analysis of variance (ANOVA) was conducted to indicate the OM differences in the four wetlands. A two-way ANOVA was conducted to determine the effects of two wetland types (constructed/natural wetlands) and two wetland habitats (invaded/normal wetland habitats) on the OM concentrations and storages for each sediment depth. Meanwhile, the contents of C and N fractions and LFOC LFON ratio of all sampling sites were also independently analyzed for the effects of two sediment depths. The above-mentioned steps were conducted in SPSS (version 21.0).

To evaluate microbial alpha diversity, we calculated the abundance-based coverage estimator (ACE) index, using an ACE calculator (The Department of Microbiology & Immunology at The University of Michigan, Detroit, MI, USA. 2009. https://www.mothur.org/wiki/Ace) [32]. Observed OTUs were counted in Mothur, and OTU-level ranked abundance curves were generated to compare the richness and evenness of OTUs among groups (Figure S1). Principal Component Analysis (PCA) of bacterial phyla was performed in Canoco (version 4.5, Wageningen University and Research Centre, Vakhn, Netherlands, 2002) [33]. All bacterial taxa at the phylum level and the ACE index were checked to remove the outliers, and all data used obeyed normal distribution. For the potential effects of alligator weed on bacterial composition, we conducted the mean value analysis of the bacterial phyla in the four wetlands, which differentiated between the invaded and normal habitats. A two-way ANOVA was conducted to determine the significant differences of bacterial taxa and ACE index between constructed/natural wetlands and between invaded/normal habitats. The Benjamini–Hochberg algorithm was processed to control false discovery rate at the significance level of 0.05 for the two-way ANOVA of bacterial phyla in R (version 3.2.2) [34]. Cluster analysis by the unweighted pair-group method with arithmetic means was conducted to the bacterial phyla among the four wetlands that differentiated between invaded and normal habitats in R (version 3.2.2). Pearson correlation analysis was also conducted to assess the potential associations between the bacterial taxa and OM fractions, using the SPSS software (version 21.0). Concretely, the p-values for the Pearson correlation analysis meant the significances between the two variables in this study indicated significant correlation, and extremely significant correlation when the p-values were less than 0.05 and 0.01, respectively.

3. Results and Discussion

3.1. Organic Matter in the Wetland Habitats

In this study, the contents and storage of the C and N fractions did not present noteworthy difference among the wetlands of XRCW, NLE, XR, and NL. This indicated that the different wetlands were not an important factor affecting the OM contents and storage (Table 1). This finding was in accordance with the reports of Cao et al. [35], who showed the similar distribution trend of organic carbon in XRCW and XR. Specifically, LFOC and HFON reached the maximal average values of 0.834, and 0.414 $g{\cdot}kg^{-1}$ in the surface sediments of NLE, LFON reached the maximal average content (0.043 $g{\cdot}kg^{-1}$) and storage (0.010 $kg{\cdot}m^{-2}$) in the surface sediments of XRCW, while HFOC contents and storage of HFOC and HFON were the greatest in NL with the average values of 15.55 $g{\cdot}kg^{-1}$, and 4.199 and 0.078 $kg{\cdot}m^{-2}$. Hogan et al. [36] indicated that natural wetlands had a higher OC concentration than the constructed wetlands. Further, Bruland and Richardson [37] showed that non-riverine organic soil flat hydrogeomorphic of natural or constructed wetlands had significantly higher OC contents. Due to the extensive flat transition area from XRCW to NL, NLE showed the highest moisture content in this study.

The distribution and storage characteristic of organic matter in the natural wetlands (XR and NL) and the constructed wetlands (XRCW and NLE) were analyzed in this study (Table 2). Results showed that contents of LFOC and LFON in the surface sediments were significantly higher in the constructed wetlands (mean values of 0.791 and 0.043 $g{\cdot}kg^{-1}$) than those in the natural wetlands (mean values of 0.500 and 0.022 $g{\cdot}kg^{-1}$), while HFs showed no significant difference. The finding indicated the sensitivity of LFOC and LFON, and the effects of wetland types were mainly on the LFs. Gao et al. [38] also showed that LFOC was the best potential indicator of the OC dynamics. Nelson et al. [39] reported that high C inputs contributed to the high LFOC contents in restored grassland. In this study, the significantly higher contents of LFOC and LFON in the constructed wetlands than in natural wetlands might result from the high plant cover and the corresponding input of LFs into sediments [12,40]. The moisture content was also remarkably higher in the constructed wetlands (76%) than in the natural wetlands (45%), the result was in accordance with our previous report [21]. It was reported that moisture contents had positive and significant relationships with the composition of clay and silt, while it had negative associations with proportion of sand [41]. Thus, the higher moisture contents in the constructed wetlands might have resulted from the higher proportion of clay and silt in the constructed wetlands (86.85%) than in the natural wetlands (79.78%) [18,19]. In addition, sampling-time difference and root activities might also be important factors [42]. The storages of LFOC, LFON, and HFON in he constructed wetlands were relatively higher but not significant. A previous study showed that storages of OC and ON had no significant difference between wetland types of bogs and fens [43], suggesting that the OM storage was relatively stable and not easily affected by different wetland types.

Table 1. The concentrations ($g \cdot kg^{-1}$) and storage ($kg \cdot m^{-2}$) of C and N fractions in the surface sediments of the four wetland areas (mean values ± standard deviation).

Parameters	XRCW	NLE	XR	NL	*p*-Values
LFOC	0.771 ± 0.28	0.834 ± 0.41	0.460 ± 0.24	0.601 ± 0.04	0.173
LFON	0.043 ± 0.021	0.042 ± 0.024	0.022 ± 0.014	0.025 ± 0.005	0.182
HFOC	13.92 ± 5.05	14.80 ± 8.70	13.27 ± 5.37	15.55 ± 8.03	0.961
HFON	0.290 ± 0.21	0.414 ± 0.32	0.115 ± 0.11	0.308 ± 0.27	0.360
$S_{(LFOC)}$	0.170 ± 0.06	0.164 ± 0.07	0.147 ± 0.09	0.168 ± 0.014	0.937
$S_{(LFON)}$	0.010 ± 0.01	0.008 ± 0.005	0.007 ± 0.005	0.0068 ± 0.003	0.798
$S_{(HFOC)}$	3.20 ± 1.14	2.75 ± 1.30	3.80 ± 1.27	4.199 ± 1.58	0.350
$S_{(HFON)}$	0.061 ± 0.06	0.071 ± 0.06	0.031 ± 0.04	0.078 ± 0.09	0.619
Moisture content (%)	0.682 ± 0.17	0.925 ± 0.55	0.441 ± 0.10	0.464 ± 0.14	0.054
Bulk density ($g \cdot cm^{-3}$)	0.979 ± 0.16	0.852 ± 0.29	1.146 ± 0.16	1.125 ± 0.18	0.087

$S_{(LFOC)}$ means the storage of light fraction organic carbon (LFOC) in wetland sediments of 0–15 cm depth. The *p* values indicated the distribution differences among the four wetlands by one-way ANOVA.

The areas invaded by alligator weed showed significantly higher moisture contents than the normal wetland habitats, root growth and activities might be the main driving factors [42]. While Hossler [44] showed that the newly created (constructed) wetland had a higher soil bulk density, less moisture content, plant biomass, or SOC than the natural wetlands. The opposite results might indicate that wetland age was also an important factor affecting the soil physical properties and the OM accumulation level [45]. In the surface sediments, the contents of the C and N fractions in invaded habitats were prominently higher than those in normal wetland habitat. This was in accordance with the reports of Yang et al. [46] and Zhang et al. [47], who observed the increases in the contents of OC and ON under *Spartina alterniflora* invasion. On average, in this study, the LFOC, LFON, HFOC, and HFON of invaded habitats were 0.265 $g \cdot kg^{-1}$, 0.029 $g \cdot kg^{-1}$, 5.27 $g \cdot kg^{-1}$, and 0.278 $g \cdot kg^{-1}$ higher than those in normal habitats. Zhang et al. [47] showed that *S. alterniflora* invasion promoted the increase of OC and ON storage for 3.67–4.90 $g\ C \cdot kg^{-1}$ and 0.307–0.391 $g\ N \cdot kg^{-1}$, respectively. The increasing level of ON was higher than that in the invaded sediments of this study, while the increase in OC was less than that in our results. Moreover, Singh et al. [48] indicated higher pH values in the invaded soil than in the non-invaded soil and Dlamini et al. [49] showed a significantly higher depletion of OC in acidic soil ($pH < 5$) than in soil with high pH. Therefore, the high accumulation of OC in the invaded areas might profit from the potential high pH values in this study, relative works need to be done to confirm the deduction. Furthermore, the ON storages were also much higher in the invaded habitats than in the normal habitats (Table 2). However, the invasion of *S. alterniflora* and *Lythrum salicaria* dramatically stimulated N mineralization and nitrification, by adding N input, thus the storage of ON was unchanged [50–52]. Therefore, this study demonstrated that alligator weed invasion could be beneficial to the net storage of N fractions. It was reported that the deposition of C could be inhibited by a low N content, given that the denitrification process was closely associated with that of OC decomposition [53,54]. Therefore, an increase of N input was considered favorable for the deposition of OC.

The contents of LFOC and LFON were 0.721 and 0.038 g·kg^{-1} in the surface sediments (0–15 cm), significantly higher than they were in the subsurface (15–25 cm), at 0.511 and 0.022 g·kg^{-1} respectively (Figure 2). The accumulation and mineralization of LFs were dynamic and sensitive, and easily affected by climate change, wetland hydrology, and soil physical components [55]. Zhang et al. [20] and Zhang et al. [56] also reported that the OC contents significantly decreased in the soil profiles of reed wetland, paddy field, fens, and humus marsh. As the decomposable fraction, LFs were remarkably decreased in the wetland profile [21,57], which could be caused by their physicochemical lability. In the surface sediments, the effects of wetland types or wetland habitats on the LFs contents were significant, while the effects were mostly insignificant in the subsurface sediments, indicating that the C and N fractions in the surface sediments were easily changed and affected by environmental factors. LFOC:LFON values showed no difference between natural and constructed wetlands in surface sediments, while it was notably higher in natural wetlands than in constructed wetland in the subsurface sediments. Wang et al. [43] indicated the interacting effects on OC and ON but not on the C:N ratio, indicating that the OC and ON might have different distribution patterns affected by the wetland types and depths. In this study, the changes of LFOC:LFON values between the two depths were mostly caused by the higher decreasing rate of LFON (41.66%) than that of LFOC (29.09%) from the surface to the subsurface sediments, which further indicated the sensitivity of LFON to the sampling depths.

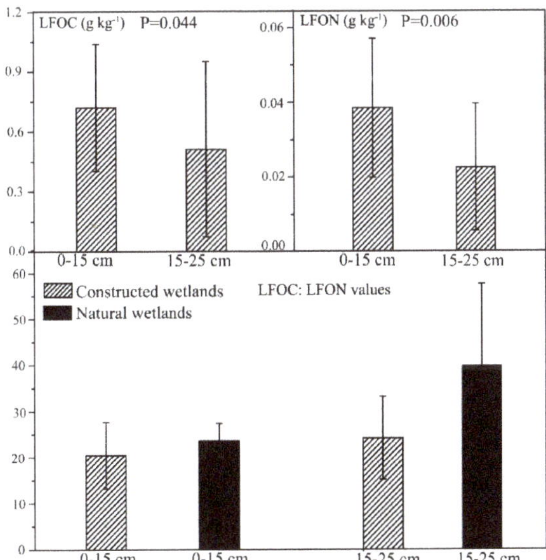

Figure 2. The distribution characteristic of LFOC, LFON, and the LFOC:LFON values in the two sediment depths (0–15 cm and 15–25 cm). The error bars indicate the standard deviation. One-way ANOVA of two depths was conducted on LFOC and LFON, and the mean value analysis was also implemented to the LFOC:LFON ratio.

Table 2. The average concentrations ($g \cdot kg^{-1}$) and storage ($kg \cdot m^{-2}$) of C and N fractions (mean values ± standard deviation) and the two-way ANOVA (constructed/natural wetlands and invaded/normal habitats) for each depths.

Parameters (0–15 cm)	**LFOC**	**LFON**	**HFOC**	**HFON**	**S(LFOC)**	**S(LFON)**	**S(HFOC)**	**S(HFON)**
Constructed wetlands	0.791 ± 0.32	0.043 ± 0.022	14.20 ± 6.23	0.313 ± 0.24	0.168 ± 0.07	0.010 ± 0.008	3.058 ± 1.18	0.064 ± 0.056
Natural wetlands	0.500 ± 0.21	0.022 ± 0.011	13.92 ± 5.59	0.175 ± 0.14	0.153 ± 0.07	0.007 ± 0.004	3.913 ± 1.23	0.044 ± 0.055
p values	0.012	0.004	0.912	0.146	0.633	0.255	0.138	0.431
Invaded habitats	0.885 ± 0.29	0.056 ± 0.019	17.40 ± 5.31	0.464 ± 0.26	0.174 ± 0.08	0.012 ± 0.011	3.610 ± 1.24	0.087 ± 0.060
Normal habitats	0.620 ± 0.29	0.027 ± 0.015	12.13 ± 5.598	0.186 ± 0.18	0.159 ± 0.06	0.007 ± 0.003	3.053 ± 1.21	0.042 ± 0.046
p values	0.027	0.000	0.019	0.014	0.585	0.152	0.249	0.046
Parameters (15–25 cm)	LFOC	LFON	HFOC	HFON	S(LFOC)	S(LFON)	S(HFOC)	S(HFON)
Constructed wetlands	0.572 ± 0.45	0.026 ± 0.018	15.27 ± 6.36	0.295 ± 0.18				
Natural wetlands	0.245 ± 0.19	0.006 ± 0.002	12.11 ± 4.01	0.180 ± 0.15				
p values	0.137	**0.042**	0.23	0.169				
Invaded habitats	0.688 ± 0.58	0.030 ± 0.02	16.29 ± 6.91	0.328 ± 0.15				
Normal habitats	0.390 ± 0.26	0.017 ± 0.01	13.42 ± 5.23	0.232 ± 0.18				
p values	0.082	0.079	0.214	0.202				

It showed the significant and extremely significant difference, when *p* values were less than 0.05 and 0.01, respectively (in bold). $S_{(LFOC)}$ means the storage of LFOC.

3.2. Bacterial Distribution and Composition

The PCA was conducted and the first two axes explained 90.3% of the sample variation (Figure 3). Thereinto, a variance of 81.8% was attributed to principal component 1, and principal component 2 captured 8.5% of the variance. The samples, from the constructed or the natural wetland, and from the invaded area or the normal wetland habitats, were mostly decentralized and distributed, and showed no obvious aggregation. As a whole, this indicated that the factors of invaded/normal wetland habitats and constructed/natural wetlands might not cause notable changes to the sediment microflora.

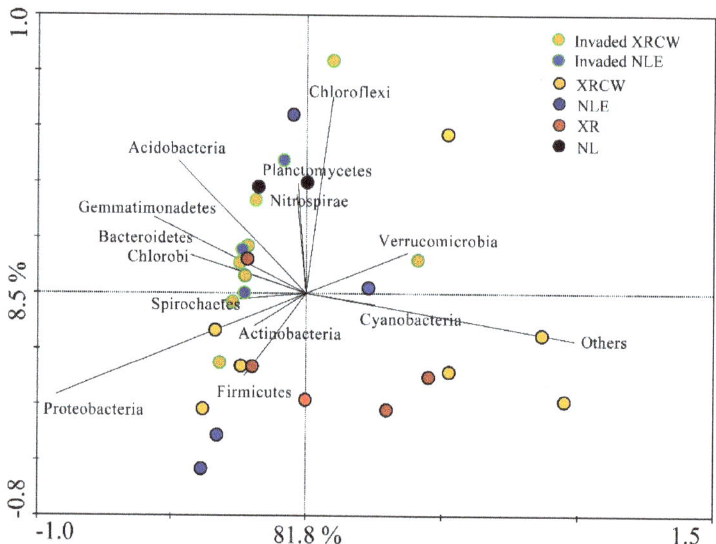

Figure 3. Principal component analysis (PCA) based on the sampling sites and bacterial phyla composition.

The composition pattern of the bacterial phyla was analyzed and the cluster analysis of the four wetlands that differentiated between invaded and normal habitats was implemented (Figure 4). Cluster analysis showed that the invaded areas were gathered first, and then clustered with NLE, which followed by NL, XRCW, and XR, accordingly. The finding suggested that the alligator weed invasion was a dominant factor that affected the bacterial phyla composition, while the effect of wetland types was not obvious, due to the scattered cluster between natural and constructed wetlands.

Specifically, Proteobacteria (55.94%), Bacteroidetes (5.74%), Acidobacteria (6.66%), and Chloroflexi (4.67%) were the dominant bacterial phyla in the studied areas. Among the dominant phyla, Firmicutes was significantly higher in the constructed wetlands (2.01%) than in the natural wetlands (1.43%; adjusted p value of 0.044), while distribution of Gemmatimonadetes was the opposite (adjusted p value of 0.026). Proportion of Cyanobacteria was also very high in the natural wetlands (0.55%) than in the constructed wetlands (0.33%). The results were similar with our previous study on these wetlands, indicating the relatively stable composition of bacterial phyla [21]. Moreover, Adrados et al. [58] showed the significant differences of Bacteroidetes proportions among different wetland units, and it was mostly associated with the degradation of high molecular weight compounds and complex organic particles, such as cellulose and lignin [59,60]. The insignificant distribution of Bacteroidetes in this study indicated its stability under the effects of wetland types and the invaded/normal habitats. Cyanobacteria can reportedly degrade pyrene and other complex organics, and it was very abundant in oil-polluted sediments [61]. Thus, the high proportion of Cyanobacteria in natural wetlands might contribute to the degradation of the C and N fractions, which result in the corresponding low contents, especially in XR. In addition, as a ubiquitous colonizer, Cyanobacteria prefers a high NO_3^--N and NH_4^+-N environment, thus, the higher proportion of Cyanobacteria in natural wetlands than in constructed wetlands might indicate potential eutrophication, especially in NL. Firmicutes was mainly identified as a denitrifier [62], including abundant bacterial taxa that reduce the nitrate into ammonia in anaerobic environment and improve sewage purification in constructed wetlands. Therefore, the high Firmicutes in constructed wetlands and high Cyanobacteria in natural wetlands promote the metabolism of NO_3^- and NH_4^+ into ON and ammonia, respectively. Nitrospirae is one typical nitrifying bacteria that oxidize NH_3 and NO_2^- into NO_3^- [63], it showed no significant difference between the two wetland types or between the invaded and normal wetland habitats, with a mean proportion of 3.58% in this study.

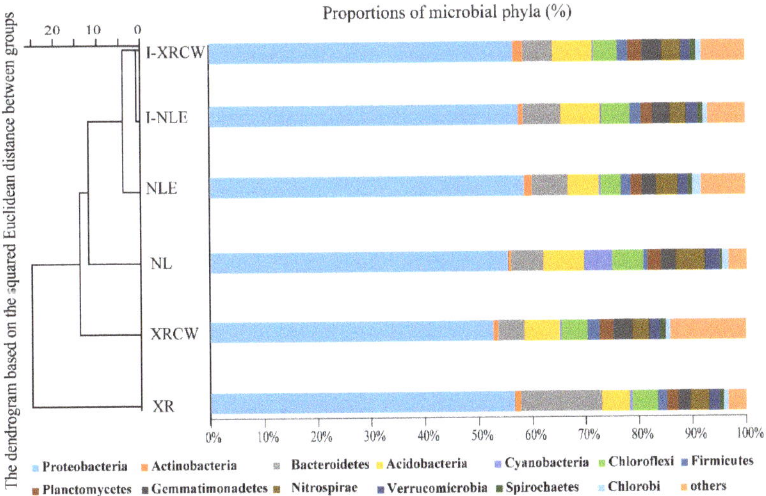

Figure 4. The cluster analysis of the invaded and normal wetlands and the composition characteristic of the dominant bacterial phyla. I-XRCW/I-NLE—Invaded Xinxue River Constructed Wetland/Invaded Nansi Lake estuary.

In the level of bacterial phyla taxa, Acidobacteria, Actinobacteria, and Gemmatimonadetes were significantly higher in the invaded areas than in the normal wetland habitats (adjusted p values were 0.028, 0.041, and 0.045, respectively). It was reported that Acidobacteria and Actinobacteria were important microbial components that closely associate with the decomposition of biological residues and carbon mineralization, respectively [21,64]; thus, their high proportions in the invaded areas might suggest the high efficiency of carbon sequestration and mineralization. The ACE index was much higher in both the invaded areas (6332.8) and the constructed wetlands (6170.0) than in normal or natural wetland conditions (5838.7 and 5574.0), with p values of 0.128 and 0.158, respectively. This suggested that the invaded and constructed wetland habitats were beneficial to the increase in bacterial diversity. Batten et al. [15] showed that the invasion of *C. solstitialis* and *A. triuncialis* significantly changed the soil bacterial composition pattern, and the newly invaded areas had a microbial composition similar to the original native soils. Alligator weed invasion increased bacterial diversity and the abundance of Acidobacteria, Actinobacteria, and Gemmatimonadetes. At the class level, the α-, β-, γ-, and δ-*proteobacteria* accounted for 9.22%, 14.61%, 12.23%, and 16.82% of the microbial communities, respectively, which were the dominant classes in this study. The Pearson correlation analysis showed that proportions of Acidobacteria were significantly and positively associated with the concentrations of LFON ($p = 0.425$, $R^2 = 0.187$) and HFON ($p = 0.474$, $R^2 = 0.224$). This indicated that the Acidobacteria activities might promote the ON accumulation.

4. Conclusions

In this study, the contents of LFs significantly differed under the effects of wetland types and alligator weed invasion. The contents of LFs were also sensitive and easily affected in the surface sediments, as compared to the subsurface sediments. The storages of OM were very stable. Proteobacteria, Bacteroidetes, Acidobacteria, and Chloroflexi were the dominant bacterial phyla and abundance of Acidobacteria, Actinobacteria, and Gemmatimonadetes were significantly higher in the invaded habitats than in normal habitats. The high ACE index in the constructed wetlands and invaded areas suggested the corresponding high bacterial diversity. In addition, Acidobacteria had significant and positive effects on ON accumulation. This study contributed to the understanding of the effects of alligator weed invasion on the composition characteristics of organic matter and bacterial communities in wetland, which is important to predict the function of wetlands, under plant invasions.

Supplementary Materials: The following are available online at http://www.mdpi.com/2076-3417/10/16/5571/s1, Figure S1: The OTU-level ranked abundance curves of the 29 samples in this study.

Author Contributions: Conceptualization, Q.C. and J.L.; methodology, W.M.; software, H.Z.; validation, R.W. and J.L.; formal analysis, W.M.; investigation, Q.C.; data curation, Q.C.; writing—original draft preparation, Q.C. and J.L.; writing—review and editing, Q.C., H.Z., and J.L.; visualization, H.Z.; supervision, R.W.; funding acquisition, Q.C. and J.L. All authors have read and agreed to the published version of the manuscript.

Funding: This work was supported by the National Key R&D Program of China (No. 2017YFC0505905), the Natural Science Foundation of Shandong Province, China (ZR2017MC013), and the Science Foundation of Shandong Jianzhu University (Grant No. X18047ZX).

Acknowledgments: We would like to thank Editage (www.editage.com) for English language editing.

Conflicts of Interest: The authors declare no conflict of interest.

References

1. Bassett, I.; Paynter, Q.; Beggs, J.; Preston, C.; Watts, J.H.; Crossman, N.D. Alligator weed (*Alternanthera philoxeroides*) invasion affects decomposition rates in a northern New Zealand lake. In Proceedings of the Managing Weeds in a Changing Climate, Adelaide, South Australia, 24–28 September 2006; pp. 776–779.
2. Palihakkara, C.R.; Dassanayake, S.; Jayawardena, C.; Senanayake, I.P. Floating wetland treatment of acid mine drainage using *Eichhornia crassipes* (water hyacinth). *J. Health. Pollut.* **2018**, *8*, 14–19. [CrossRef] [PubMed]
3. Hao, W.; Juli, C.; Jianqing, D. Invasion by alligator weed, *alternanthera philoxeroides*, is associated with decreased species diversity across the latitudinal gradient in China. *J. Plant Ecol.* **2015**, *3*, 311–319.
4. Spencer, N.R.; Coulson, J.R. The biological control of alligator weed, *Alternanthera philoxeroides*, in the United States of America. *Aquat. Bot.* **1976**, *2*, 177–190. [CrossRef]

5. Tao, Y.; Jiang, M. Study on anatomical structure adaptation of stem of *Alternanthera philoxeroides (Mart.) Griseb* to various water condition. *Wuhan Bot. Res.* **2003**, *22*, 65–71.
6. Chen, G.; Azkab, M.H.; Chmura, G.L.; Chen, S.; Sastrosuwondo, P.; Ma, Z.; Dharmawan, I.W.E.; Yin, X.; Chen, B. Mangroves as a major source of soil carbon storage in adjacent seagrass meadows. *Sci. Rep.-UK* **2017**, *7*, 1–10. [CrossRef]
7. Skornia, K.; Safferman, S.I.; Rodriguez-Gonzalez, L.; Ergas, S.J. Treatment of winery wastewater using bench-scale columns simulating vertical flow constructed wetlands with adsorption media. *Appl. Sci.* **2020**, *10*, 1063. [CrossRef]
8. Geddes, P.; Grancharova, T.; Kelly, J.J.; Treering, D.; Tuchman, N.C. Effects of invasive *Typha* × *glauca* on wetland nutrient pools, denitrification, and bacterial communities are influenced by time since invasion. *Aquat. Ecol.* **2014**, *48*, 247–258. [CrossRef]
9. Martina, J.P.; Hamilton, S.K.; Turetsky, M.R.; Phillippo, C.J. Organic matter stocks increase with degree of invasion in temperate inland wetlands. *Plant Soil.* **2014**, *385*, 107–123. [CrossRef]
10. Nguyen, H.T.T.; Chao, H.R.; Chen, K.C. Treatment of organic matter and tetracycline in water by using constructed wetlands and photocatalysis. *Appl. Sci.* **2019**, *9*, 2680. [CrossRef]
11. Whalen, J.K.; Bottomley, P.J.; Myrold, D.D. Carbon and nitrogen mineralization from light-and heavy-fraction additions to soil. *Soil Biol. Biochem.* **2000**, *32*, 1345–1352. [CrossRef]
12. Tan, Z.; Lal, R.; Owens, L.; Izaurralde, R.C. Distribution of light and heavy fractions of soil organic carbon as related to land use and tillage practice. *Soil Till. Res.* **2007**, *92*, 53–59. [CrossRef]
13. Liao, J.D.; Boutton, T.W.; Jastrow, J.D. Storage and dynamics of carbon and nitrogen in soil physical fractions following woody plant invasion of grassland. *Soil Biol. Biochem.* **2006**, *38*, 3184–3196. [CrossRef]
14. Bertin, C.; Yang, X.; Weston, L. The role of root exudates and allelochemicals in the rhizosphere. *Plant Soil.* **2003**, *256*, 67–83. [CrossRef]
15. Batten, K.M.; Scow, K.M.; Davies, K.F.; Harrison, S.P. Two invasive plants alter soil microbial community composition in serpentine grasslands. *Biol. Invasions.* **2006**, *8*, 217–230. [CrossRef]
16. Cao, Q.; Hui, W.; Chen, X.; Wang, R.; Jian, L. Composition and distribution of microbial communities in natural river wetlands and corresponding constructed wetlands. *Ecol. Eng.* **2017**, *98*, 40–48. [CrossRef]
17. Zhang, J.; Zhang, B.; Jing, Y.; Kang, X.; Zhang, C. Design of constructed wetland in Xinxue River Estuary into Nansi Lake in Eastern Line of South-to-North water transfer project. *China Water Wastewater* **2008**, *24*, 49–51.
18. Zhang, Y. Plant Diversity, Soil Mercury Risk and Eco-Sustainability of Xinxue River Constructed Wetland in Nansi Lake, China. Ph.D. Thesis, Shandong University, Jinan, China, 2014.
19. Liu, H. Distribution Characteristics, Bioaccumulation, and Sources of Mercury in Rice at Nansi Lake Area, Shandong Province, China. *JAPS J. Anim. Plant Sci.* **2015**, *25*, 114–121.
20. Zhang, W.J.; Xiao, H.A.; Tong, C.L.; Su, Y.R.; Xiang, W.S.; Huang, D.Y.; Syers, J.K.; Wu, J. Estimating organic carbon storage in temperate wetland profiles in Northeast China. *Geoderma* **2008**, *146*, 311–316. [CrossRef]
21. Cao, Q.; Wang, R.; Zhang, H.; Ge, X.; Liu, J. Distribution of organic carbon in the sediments of Xinxue River and the Xinxue River Constructed Wetland, China. *PLoS ONE* **2015**, *10*, e0134713. [CrossRef]
22. Moiwo, J.P.; Wahab, A.; Kangoma, E.; Blango, M.M.; Ngegba, M.P.; Suluku, R. Effect of biochar application depth on crop productivity under tropical rainfed conditions. *Appl. Sci.* **2019**, *9*, 2602. [CrossRef]
23. Bambi, P.; Rezende, R.D.S.; Feio, M.J.; Leite, G.F.M.; Alvin, E.; Quintão, J.M.B.; Araújo, F.; Júnior, J.F.G. Temporal and spatial patterns in inputs and stock of organic matter in Savannah Streams of Central Brazil. *Ecosystems* **2017**, *20*, 757–768. [CrossRef]
24. Nadzirah, K.Z.; Zainal, S.; Noriham, A.; Normah, I. Efficacy of selected purification techniques for bromelain. *Int. Food Res. J.* **2013**, *20*, 43–46.
25. Niemi, R.M.; Heiskanen, I.; Wallenius, K.; Lindström, K. Extraction and purification of DNA in rhizosphere soil samples for PCR-DGGE analysis of bacterial consortia. *J. Microbiol. Meth.* **2001**, *45*, 155–165. [CrossRef]
26. Caporaso, J.G.; Lauber, C.L.; Walters, W.A.; Berg-Lyons, D.; Lozupone, C.A.; Turnbaugh, P.J.; Fierer, N.; Knight, R. Global patterns of 16S rRNA diversity at a depth of millions of sequences per sample. *Proc. Natl. Acad. Sci. USA* **2011**, *108*, 4516–4522. [CrossRef] [PubMed]
27. Caporaso, J.G.; Kuczynski, J.; Stombaugh, J.; Bittinger, K.; Bushman, F.D.; Costello, E.K.; Fierer, N.; Peña, A.G.; Goodrich, J.K.; Gordon, J.I.; et al. QIIME allows analysis of high-throughput community sequencing data. *Nat. Methods* **2010**, *7*, 335–336. [CrossRef] [PubMed]

28. Dolgova, A.S.; Stukolova, O.A. High-fidelity PCR enzyme with DNA-binding domain facilitates de novo gene synthesis. *Biotech* **2017**, *7*, 1–6. [CrossRef]
29. Edgar, R.C. Search and clustering orders of magnitude faster than BLAST. *Bioinformatics* **2010**, *26*, 2460–2461. [CrossRef]
30. García, A.L.; Quiroga, C.P.; Atxaerandio, R.; Pérez, A.; Recio, O.G. Comparison of Mothur and Qiime for the analysis of rumen microbiota composition based on 16s rRNA amplicon sequences. *Front. Microbiol.* **2018**, *9*, 1–11.
31. DeSantis, T.Z.; Hugenholtz, P.; Larsen, N.; Rojas, M.; Brodie, E.L.; Keller, K.; Huber, T.; Dalevi, D.; Hu, P.; Andersen, G.L. Greengenes, a Chimera-Checked 16S rRNA Gene Databaseand Workbench Compatible with ARB. *Appl. Environ. Microbial.* **2006**, *72*, 5069–5072. [CrossRef]
32. Schloss, P.D. A high-throughput DNA sequence aligner for microbial ecology studies. *PLoS ONE* **2009**, *4*, e8230. [CrossRef]
33. Ramette, A. Multivariate analyses in microbial ecology. *FEMS Microbiol. Ecol.* **2007**, *62*, 142–160. [CrossRef]
34. Benjamini, Y.; Hochberg, Y. Controlling the false discovery rate: A practical and powerful approach to multiple testing. *J. R. Stat. Soc. B.* **1995**, *57*, 289–300. [CrossRef]
35. Cao, Q.; Wang, H.; Zhang, Y.; Lal, R.; Wang, R.; Ge, X.; Jian, L. Factors affecting distribution patterns of organic carbon in sediments at regional and national scales in China. *Sci. Rep.-UK* **2017**, *7*, 1–10. [CrossRef] [PubMed]
36. Hogan, D.M.; Jordan, T.E.; Walbridge, M.R. Phosphorus retention and soil organic carbon in restored and natural freshwater wetlands. *Wetlands* **2004**, *24*, 573–585. [CrossRef]
37. Bruland, G.L.; Richardson, C.J. Comparison of soil organic matter in created, restored and paired natural wetlands in North Carolina. *Wetl. Ecol. Manag.* **2006**, *14*, 245–251. [CrossRef]
38. Gao, J.; Lei, G.; Zhang, X.; Wang, G. Can δ^{13}C abundance, water-soluble carbon, and light fraction carbon be potential indicators of soil organic carbon dynamics in Zoigê Wetland? *Catena* **2014**, *119*, 21–27. [CrossRef]
39. Nelson, J.D.J.; Schoenau, J.J.; Malhi, S.S. Soil organic carbon changes and distribution in cultivated and restored grassland soils in Saskatchewan. *Nutr. Cycl. Agroecosys.* **2008**, *82*, 137–148. [CrossRef]
40. Lehmann, J.; Skjemstad, J.; Sohi, S.; Carter, J.; Barson, M.; Falloon, P.; Coleman, K.; Woodbury, P.B. Australian climate–carbon cycle feedback reduced by soil black carbon. *Nat. Geosci.* **2008**, *1*, 832–835. [CrossRef]
41. Ma, S.; Xie, Y.; Hu, H.; Ni, B. Relationship between soil water content and soil particle distribution in two kinds of typical community types of desert steppe. *Soil Water Conserv. China* **2019**, *7*, 61–65.
42. Gerke, H.H.; Kuchenbuch, R.O. Root effects on soil water and hydraulic properties. *Biologia* **2007**, *62*, 557–561. [CrossRef]
43. Wang, X.; Song, C.; Sun, X.; Wang, J.; Zhang, X.; Mao, R. Soil carbon and nitrogen across wetland types in discontinuous permafrost zone of the Xiao Xing'an Mountains, Northeastern China. *Catena* **2013**, *101*, 31–37. [CrossRef]
44. Hossler, K. Accumulation of carbon in created wetland soils and the potential to mitigate loss of natural wetland carbon-mediated functions. Ph.D. Thesis, The Ohio State University, Environmental Science, Columbus, OH, USA, 2005.
45. De Mastro, F.; Cocozza, C.; Brunetti, G.; Traversa, A. Chemical and spectroscopic investigation of different soil fractions as affected by soil management. *Appl. Sci.* **2020**, *10*, 2571. [CrossRef]
46. Yang, W.; Zhao, H.; Leng, X.; Cheng, X.; An, S. Soil organic carbon and nitrogen dynamics following *Spartina alterniflora* invasion in a coastal wetland of Eastern China. *Catena* **2017**, *156*, 28–289. [CrossRef]
47. Zhang, Y.; Ding, W.; Luo, J.; Donnison, A. Changes in soil organic carbon dynamics in an Eastern Chinese coastal wetland following invasion by a C4 plant *Spartina alterniflora*. *Soil Biol. Biochem.* **2010**, *42*, 1712–1720. [CrossRef]
48. Singh, M.R.; Theunuo, N. Variation of soil pH, moisture, organic carbon and organic matter content in the invaded and non-invaded areas of *Tithonia diversifolia* (Hemsl.) A. gray found in Nagaland, North-east India. *Eco. Env. Cons.* **2017**, *23*, 2181–2187.
49. Dhillon, J.; Del Corso, M.R.; Figueiredo, B.; Nambi, E.; Raun, W. Soil organic carbon, total nitrogen, and soil ph, in a long-term continuous winter wheat (*Triticum aestivum l.*) experiment. *Commun. Soil Sci. Plan.* **2018**, *8*, 1–11. [CrossRef]
50. Fickbohm, S.S.; Zhu, W.X. Exotic purple loosestrife invasion of native cattail freshwater wetlands: Effects on organic matter distribution and soil nitrogen cycling. *Appl. Soil Ecol.* **2006**, *32*, 123–131. [CrossRef]

51. Martini, J.; Orge, C.A.; Faria, J.L.; Pereira, M.F.R.; Soares, O.S.G.P. Catalytic advanced oxidation processes for sulfamethoxazole degradation. *Appl. Sci.* **2019**, *9*, 2652. [CrossRef]
52. Zhang, Y.; Xu, X.; Li, Y.; Huang, L.; Xie, X.; Dong, J.; Yang, S. Effects of *Spartina alterniflora* invasion and exogenous nitrogen on soil nitrogen mineralization in the coastal salt marshes. *Ecol. Eng.* **2016**, *87*, 281–287.
53. Moore, T.R.; Trofymow, J.; Prescott, C.E.; Titus, B.; Group, C.W. Nature and nurture in the dynamics of C, N and P during litter decomposition in Canadian forests. *Plant Soil* **2011**, *339*, 163–175. [CrossRef]
54. Zhao, W.; Liu, Y.; Wei, H.; Zhang, R.; Luo, G.; Hou, H.; Chen, S.; Zhang, R. NO removal by Plasma-Enhanced NH_3-SCR using methane as an assistant reduction agent at low temperature. *Appl. Sci.* **2019**, *9*, 2751. [CrossRef]
55. Huo, L.; Chen, Z.; Zou, Y.; Lu, X.; Guo, J.; Tang, X. Effect of Zoige Alpine Wetland degradation on the density and fractions of soil organic carbon. *Ecol. Eng.* **2013**, *51*, 287–295. [CrossRef]
56. Zhang, W.; Peng, P.; Tong, C.; Wang, X.; Wu, J. Characteristics of distribution and composition of organic carbon in Dongting Lake Floodplain. *Environ. Sci.* **2005**, *26*, 56–60.
57. Six, J.; Merckx, R.; Kimpe, K.; Paustian, K.; Elliott, E.T. A re-evaluation of the enriched labile soil organic matter fraction. *Eur. J. Soil Sci.* **2000**, *51*, 283–293. [CrossRef]
58. Adrados, B.; Sánchez, O.; Arias, C.A.; Becares, E.; Garrido, L.; Mas, J.; Brix, H.; Morató, J. Microbial communities from different types of natural wastewater treatment systems: Vertical and horizontal flow constructed wetlands and biofilters. *Water Res.* **2014**, *55*, 304–312. [CrossRef]
59. Kirchman, D. The ecology of Cytophaga-Flavobacteria in aquatic environments. *FEMS Microbiol. Ecol.* **2002**, *39*, 91–100. [CrossRef]
60. Dorador, C.; Meneses, D.; Urtuvia, V.; Demergasso, C.; Vila, I.; Witzel, K.P.; Imhoff, J.F. Diversity of Bacteroidetes in high altitude saline evaporitic basins in northern Chile. *J. Geophys. Res. Biogeoences* **2015**, *114*, 65.
61. Yan, Z.; Jiang, H.; Li, X.; Shi, Y. Accelerated removal of pyrene and benzo[a]pyrene in freshwater sediments with amendment of cyanobacteria-derived organic matter. *J. Hazard. Mater.* **2014**, *272*, 66–74. [CrossRef]
62. Liu, X.; Gao, C.; Zhang, A.; Jin, P.; Wang, L.; Feng, L. The nos gene cluster from Gram-positive bacterium *Geobacillus thermodenitrificans* NG80-2 and functional characterization of the recombinant NosZ. *FEMS Microbiol. Lett.* **2008**, *289*, 46–52. [CrossRef]
63. Fang, J.; Zhao, R.; Cao, Q.; Quan, Q.; Sun, R.; Liu, J. Effects of emergent aquatic plants on nitrogen transformation processes and related microorganisms in a constructed wetland in Northern China. *Plant Soil* **2019**, *443*, 473–492. [CrossRef]
64. De Vrieze, J.; Saunders, A.M.; He, Y.; Fang, J.; Nielsen, P.H.; Verstraete, W.; Boon, N. Ammonia and temperature determine potential clustering in the anaerobic digestion microbiome. *Water Res.* **2015**, *75*, 312–323.

MDPI
St. Alban-Anlage 66
4052 Basel
Switzerland
Tel. +41 61 683 77 34
Fax +41 61 302 89 18
www.mdpi.com

Applied Sciences Editorial Office
E-mail: applsci@mdpi.com
www.mdpi.com/journal/applsci

www.ingramcontent.com/pod-product-compliance
Lightning Source LLC
LaVergne TN
LVHW070612170726
843515LV00004B/56

* 9 7 8 3 0 3 6 5 2 4 8 4 9 *